EX Libris

THIS BOOK DONATED
FOR
Felicia Wedlowe

Observing EARTH SATELLITES

Observing EARTH SATELLITES

Desmond King-Hele

VNR VAN NOSTRAND REINHOLD COMPANY
NEW YORK   CINCINNATI   TORONTO   LONDON   MELBOURNE

Library of Congress Catalog Card Number 82-20083
ISBN 0-442-24877-6

Printed in Hong Kong

Published by Van Nostrand Reinhold Company Inc.
135 West 50th Street
New York, New York 10020

16 15 14 13 12 11 10 9 8 7 6 5 4 3 2 1

Library of Congress Cataloging in Publication Data
King-Hele, Desmond, 1927–
    Observing earth satellites.

    Bibliography: p.
    Includes index.
    1. Artificial satellites – Observers' manuals.
I. Title.
TL796.8.K47   1983   629.43'7   82-20083
ISBN 0-442-24877-6

Contents

Preface

The aim of this book is to provide a readable guide to the methods of tracking the 5000 or so artificial satellites in orbit about the Earth, with particular emphasis on do-it-yourself visual observing. About 200 of the satellites are visible to the naked eye when the conditions are right, and about 2000 can be seen with 7×50 binoculars. Watching satellites with binoculars, and timing them by stopwatch as they pass known stars, has become a valuable scientific endeavour, because the orbits determined from these and other observations can be analysed to study the shape of the Earth and the curious behaviour of its upper atmosphere. To determine accurate orbits demands observations from as many countries as possible, so satellite observing has a strongly international flavour. I hope the book will encourage more people, in many countries, to take up the challenging but rewarding art of visual observing.

My first book about observing Earth satellites was written in 1966 when the Space Age was only nine years old. Now the Space-child has matured and changed, though a few of the former features remain. The same applies to the book. In the new situation of today I have found it necessary to rewrite the book completely, though I have utilized some basic material from the original, especially among the illustrations and in Chapter 1.

As the title indicates, I do not discuss spacecraft that fly away from the Earth to the Moon or other planets. Nor do I mention observing the Moon itself, the largest of Earth satellites. This glaring omission is natural, because lunar observation is quite a different story.

I should like to pay tribute to the two pioneers whose insight and genuine kindness created the international framework for satellite observing in the early years of the Space Age. I refer to Dr Alla G. Massevitch, Vice-President of the Astronomical Council of the USSR Academy of Sciences, and Dr Fred L. Whipple, Director of the Smithsonian Astrophysical Observatory, Cambridge, Massachusetts, from 1955 to 1973. Without their efforts, international satellite observing would never have flourished as it has.

D.K.
Farnham, 1982

Note: *Throughout the book distances are in kilometres (km), and lengths in metres (m) or occasionally centimetres (cm). Masses are in kilograms (kg) or occasionally tons, where 1 ton may be taken as 1000 kg. Time appears in its familiar mixture of units – years, months, days, hours (h), minutes (min) and seconds (s). Angles are nearly always given in degrees (°).*

1
Night Lights

'The night is fine', The Walrus said.
'Do you admire the view?'

Lewis Carroll, *The Walrus and the Carpenter* (1872)

If the night is fine and dark, and not too cold, why not go out and sit in a deck-chair, and look up at the sky? When your eyes have adapted to the dark, you'll see about a thousand stars set in the black backcloth of the heavens. You'll see the whitish smudges of the Milky Way and perhaps a planet or two — and probably also a glow in the sky from the lights of the town you live in or near. But even if you are right out in the country on a moonless night, there's a faint glow from space dust sparkling in the sunlight, like specks of dust dancing on a shaft of sunlight in a darkened room. The atmosphere exudes light too, the 'night airglow', which is almost as strong as starlight but difficult to perceive: it is created by millions of atoms in the upper atmosphere excited by sunlight during the day and returning to normal by each emitting a flash of light.

The thousand stars form a fixed pattern in the sky, and give a comfortable impression of stability to the inhabitants of unstable Earth. But if you are watching the sky carefully from your deck-chair, and not just casually glancing up, you'll quite soon see some bright and transient intruders. Of these, the fastest movers are the unpredictable meteors, or shooting stars, which sometimes flash past so rapidly that you see only a streak. On most dark nights you should spot a meteor every ten minutes or so: it is probably a little fluffy grain, weaker than a petal of blossom blown from a flowering cherry tree, and it may have been travelling through space for millions of years before chance brings it into the Earth's atmosphere, where it glows white-hot and vaporizes. These 'streakers of the sky' arrive at speeds averaging about 40 kilometres per second, and their energy of motion is converted to light as they burn up at a height near 90 km. The meteor ploughs its silent furrow and is gone for ever, merely adding a little to the household dust, unless it is one of those rare

1

solid and gigantic specimens weighing tons (or hundreds of tons). These earn the name of 'meteorites' because they partially survive their fiery passage through the atmosphere: their fragments fall smoking from the sky, to the astonishment of anyone who happens to be nearby.

Artificial satellites, the second group of intruders, are very different from the meteors. They are solid structures, often of strong metal. They travel much more slowly than meteors, usually at between 7 and 8 km per second. And they are seen because they shine in the sunlight as they cross the dark sky. A satellite looks like a star that has taken leave of its senses and decided to move off to another part of the sky, like Edward Lear's 'Dong with a luminous nose':

> Then, through the vast and gloomy dark,
> There moves what seems a fiery spark,
> A lonely spark with silvery rays
> Piercing the coal-black night.

Satellites, however, are not luminous, and are only seen in the coal-black night sky because they are being illuminated by the Sun. A satellite entering the Earth's shadow immediately vanishes from view and pursues its path unseen until it emerges from eclipse again. If you look for satellites at midnight in midwinter, you won't see any, because the Earth's shadow extends for several thousand kilometres above you, and any satellites higher than that will be too faint to see. But if you go out and sit in your deck-chair an hour or two after sunset, as soon as the sky is dark, you should not have to wait more than ten minutes before you see one of the 5000 satellites now in orbit about the Earth. Of course, most of the 5000 are quite small satellites, or mere debris, and are too faint to be seen with the naked eye. But about 200 are large enough and low enough to be seen — they are mostly 6 metres or more in length and have orbits that come within 600 km of the Earth's surface. Many are as bright as the Pole Star when they cross the sky, and a few are sometimes brighter than Sirius, the brightest of the stars. They may move as fast as a high-flying airliner and are easily confused with aircraft lights. A satellite in a low orbit (at, say, 300 km height) takes about three minutes to cross the sky from horizon to horizon, but higher (and fainter) satellites can take up to half an hour or more.

Fig 1 shows the track of a satellite as recorded by a camera with a long exposure time. The picture is slightly misleading, because you really see the satellite as a point not a line: but recording it as a point is equally misleading, because it then looks like a star, whereas in reality you can see it moving.

There are many other moving lights in the sky to confuse a beginner in satellite observing, but you soon learn to recognize and ignore them — unless you are more interested in mysteries than satellites. Aircraft and helicopter lights are the most common, and weather conditions are often such that these aircraft can be seen but not heard. The other sources of lights at night are as varied as the curious human

Fig 1 *Track of Sputnik 2 over Dunfermline Abbey on 21 January 1958. The short streaks are the trails of stars during the 5-minute exposure.*

activities that produce them: weather balloons, kites and army exercises are among the most common. The possible natural phenomena are just as varied. Perhaps the most spectacular is the aurora, with its shimmering silver curtains. More static, but no less striking, are the smooth noctilucent clouds shining in the sunlight at 80 km height, and the vast variety of haloes and mock suns and moons produced by cirrus cloud. Some of these objects are sometimes called UFOs: but during many hundred hours of scanning the night sky with powerful binoculars through twenty years of satellite observing, I have never seen anything unidentifiable. So I am not much impressed by UFO reports, and I have found that those who report such sightings don't really want an identification: they prefer the sight to remain unexplained, a unique something that happened just to them, a godlike experience in a godless world.

Satellite observers do not expect anything godlike and, rather like naturalists observing the wild life of hedgerow and forest, they observe and study the wild life of the sky, learning to distinguish between different species of satellites and make

3

accurate observations of their movements and oddities of behaviour. Satellite observers are very fortunate in being able to observe at all: they have benefited from five lucky chances.

To begin with, it is quite surprising that so many satellites and rockets are bright enough to be visible to the eye. For this the observer has to thank the darkness of the night sky, the brightness of the Sun, and the inefficiency of our technology, which leads to rather large satellites and even larger rockets to launch them.

Secondly, the satellites move at about the right speed for the eye, not so slow as to be boring, but not so fast as to be gone in a flash like a meteor. They are fair game for an observer armed with binoculars and stopwatch. Satellite observing was certainly not one of the evolutionary forces that moulded the development of the eye, but the eyes we have are well adapted to satellite observing, as well as to driving cars and other unscheduled activities.

The next piece of luck is the Earth's atmosphere, which is so dense that it prevents satellites coming below a height of 100 km without burning up. So they cannot hurtle past at a height of 10 km, to be seen only by a fortunate few observers near the track, leaving all the others to gnash their teeth.

Lucky accident number four is the strong pull of gravity, which ensures that a close satellite completes its orbit within two hours. This is about the same time as the Earth's shadow takes to creep up to satellite height. So, if the orbit is well placed for observing, you should see the satellite at least once each evening, instead of finding that it is skulking on the other side of the world during the whole of the two hours.

The final slice of luck is the Earth's slow rotation rate. If it were spinning twice as fast, the shadow height would rise to satellite height within an hour or so of sunset, perhaps before the satellite had the chance to appear.

Fortified by these five fortunate facts, the ancient art of sky-watching has recently enjoyed a great revival as keen satellite observers keep track of their flock of 5000. The Stone Age sky-watchers left a lasting monument in their megalithic observatories. The Space Age sky-watchers already have their monument too — the new scientific knowledge about the shape of the Earth and the behaviour of the upper atmosphere deduced from analysis of orbits derived from their meticulous observations.

2
Up and
into Orbit

Bend, as they journey with projectile force,
In bright ellipses their reluctant course.

Erasmus Darwin, *The Botanic Garden* (1792)

Animals need oxygen to breathe, cars need fuel to keep going, but satellites will go on running round their 'bright ellipses' for thousands of years without any fuel or servicing, provided they are high enough to be free of air drag, which will otherwise whittle away their energy. The future of a satellite is determined by the magnitude and direction of its velocity at launch (unless it is among the few that have engines for manoeuvring in orbit). A satellite high enough to avoid air drag almost deserves to be called 'a creature that moves in determinate grooves', because its future course for fifty years or so can be quite well predicted. But any satellite that ventures below a height of about 2000 km subjects itself to appreciable air drag, and its future is very difficult to predict accurately, because the upper atmosphere is so extremely variable.

In this chapter I shall discuss the essentials of a satellite's life in orbit, describing its projection into orbit, taking a quick look at some of the launching rockets and then following it through life to its fiery end as it burns up in the atmosphere.

The term 'Earth satellite' includes not only instrumented spacecraft but also the final-stage rockets that nearly always accompany them into orbit, and any fragments such as cast-off nose cones or debris from explosions. Their intended functions are of little concern to the observer, who sees them all as mere points of light; and the rockets, which are often brighter than the instrumented spacecraft, may be preferred for observing.

The initial send-off

The best way to visualize the launching of a satellite is to think of a very lofty mountain, say 100 km high, with a pointed peak, and to assume for simplicity that the Earth is spherical and non-rotating. Then imagine that you climb to the summit and, inhaling strength with the mountain air, throw a stone horizontally at a speed of nearly 8 km per second. Since gravity is still almost as strong as at sea level, this stone tends to fall at much the same rate as one thrown normally at ground level: in fact, after being in flight for one second it falls about 5 m, and if the Earth were flat the stone would be 5 m nearer the surface. But the Earth is spherical (almost), not flat, and in a horizontal distance of 8 km the Earth beneath also falls away about 5 km. So

Fig 2(a) *Stone thrown at nearly 8 km per second falls 5 m while travelling 8 km and remains at the same height above the Earth.*

Fig 2(b) *Stone thrown at about 10 km per second falls 5 m while travelling 10 km and increases its height by 3 m. (Not to scale.)*

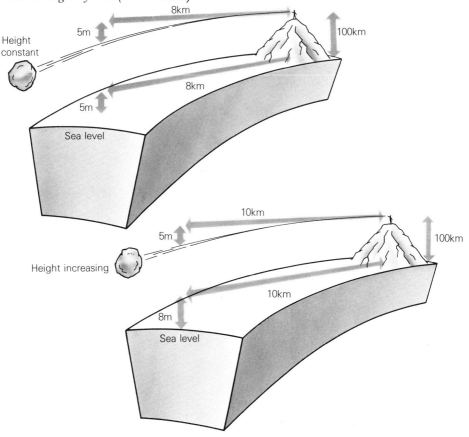

6

the stone maintains its height above the Earth, and becomes a satellite in a very nearly circular orbit at a height of 100 km, as shown in Fig 2(a), if we ignore the effect of air drag.

A launching rocket must do just what you have imagined yourself doing — lift the satellite to orbital height and send it off horizontally at that height with a speed of nearly 8 km per second (a more exact figure for a circular orbit at a height of 100 km is 7.8 km per second).

Now imagine that, going from strength to strength, you throw the stone horizontally from the mountain peak at an even faster speed, 10 km per second. After being in flight for one second, it falls about 5 m, as before. But now it has covered 10 km instead of 8, and the Earth beneath has fallen away by *more* than 5 m (actually 8 m), as shown in Fig 2(b). So the stone, despite its fall of 5 m, is now about 3 m higher: it is flying out to greater heights, and pursuing an elliptic orbit instead of a circular one.

If you are wise, you will not stand still on the mountain peak, lost in admiration at your feat. If you do, the two stones you dispatched with such *éclat* will, with even greater *éclat*, hit you in the back of the neck, the first after about 87 minutes and the second (assuming you dodged the first) after about 6 hours. Once a satellite has been injected into orbit, whether by a launching rocket or by your strong right arm, it always returns to the injection point (if we ignore some small perturbing forces). Satellites are the long-range boomerangs *par excellence*. By throwing the stone harder, you may increase its height at the other side of the world, but it still comes back to the same height at which it was hurled into orbit on your side of the world, unless you throw it so fast that it escapes from the Earth altogether.

So far we have conveniently ignored the Earth's daily rotation, which carries the mountain peak round in a west-to-east direction with the Earth, at a rate of 0.5 km per second on the equator, or 0.3 km per second at latitude 50°. Since the satellite has to attain a speed of 7.8 km per second with respect to the Earth's centre to achieve a circular orbit, the Earth's rotation provides a useful initial boost: if you launch the satellite eastwards, you only need summon up the strength to throw it at 7.3 km per second (if the mountain is on the equator), the other 0.5 km per second being supplied by the Earth's rotation. Most launching organizations are glad to accept this bonus, which allows them to reduce the size and cost of the launching rocket. So you find that most satellites travel from west-to-east rather than vice versa, though some of course need to fly over the polar regions, and cannot take advantage of the Earth's rotation.

Rocket launchers

Climbing mountain peaks to hurl satellites into orbit is impracticable, and in reality the job is done by rockets. These were first used by the Chinese in the thirteenth century and the earliest attempt at a space launching by rocket is said to have been made nearly five hundred years ago by Wan-Hu. Apparently he sat in a bamboo chair with forty-seven rockets strapped on: they were ignited simultaneously and Wan-Hu took off to become one who was never seen again, the first space martyr.

Modern rockets are more advanced in design, though not always exempt from disaster. Most of the larger satellite launching rockets have used two liquid propellants, a fuel (such as kerosene) and an oxidant (such as liquid oxygen), which are injected into a specially cooled combustion chamber, where they ignite and stream out of a nozzle as a supersonic jet of gases which thrusts the rocket forward. But many small satellites have been launched by solid-propellant rockets, burning more slowly than firework rockets but similar in principle, and having the advantage that they are much cheaper than the liquid-propellant motors.

The function of a satellite launching rocket is to give the satellite a velocity of nearly 8 km per second, and anything less is useless because the intended satellite will merely fall back into the atmosphere. The maximum velocity increase produced

Fig 3 *Lift-off of the British Black Arrow satellite launcher, which placed the Prospero satellite in orbit on 28 October 1971.*

by any rocket is equal to the velocity of the exit jet, usually about 2 km per second, multiplied by the (natural) logarithm of the 'mass ratio', that is, the initial mass divided by the final mass (when all the fuel is burnt). It is very difficult to reduce the weight of structure and motors (plus the satellite) to less than $10-15\%$ of the total weight, so the mass ratio rarely exceeds 8, and, as $\log_e 8 = 2$, a single stage rocket propulsion cannot provide a velocity increase greater than about 4 km per second. (In practice it is often less because of losses due to atmospheric drag and the 'gravity loss' arising from the low acceleration initially.) That is why at least two stages of propulsion are needed to put a satellite into orbit, and a 100 kg satellite usually needs a rocket weighing about 20 tons to launch it.

A good example of a small liquid-fuel rocket launcher was the British Black Arrow (Fig 3), developed from the very successful Black Knight high-altitude rocket. Black Arrow was a three-stage launcher, using kerosene and hydrogen peroxide as the propellants in the first two stages and a solid-propellant third stage. Black Arrow weighed 18 tons at launch, stood 13 m high and had a diameter of 2 m. The first-stage thrust was about 25% greater than the initial weight — a typical thrust/weight ratio for liquid rockets, which therefore rise slowly from their launch pads, instead of whizzing off like a firework rocket. The Black Arrow project was cancelled just before the successful launch of the 66-kg Prospero satellite from Woomera, Australia, on 28 October 1971. Prospero, and the empty third-stage rocket, entered an orbit with a minimum height of 550 km and maximum height of 1590 km. If not interfered with, they should remain in orbit for more than a hundred years.

Another example of a liquid-fuel rocket launcher, at the opposite extreme in size from Black Arrow, was the United States Saturn 5 rocket used for launching the Apollo spacecraft to the Moon in the late 1960s and early 1970s. The Saturn weighed 2900 tons at launch, stood 111 m high and had a maximum diameter of 10 m. These Saturn 5 launches were immensely impressive to watch, but, strangely enough, they have left nothing in orbit that can be seen by observers at latitudes near $50°$. A few objects from Saturn launches are still in orbit near the equator, but Skylab 1, which came down in 1979, was the last relic at $50°$ latitude.

Most people would say that the Black Arrow and Saturn 5 launchers, though so different in size, were 'conventional' in design. They are what you expect a rocket launcher to be: a large cylinder with smaller ones on top. But 'conventional' appearance should be decided by what the majority look like, and the majority of the launchings (more than 60%) have been Russian; of these the majority, from Sputnik 1 onwards, have been made by variants of a standard launch vehicle of quite different design. This launcher has four nearly conical liquid-fuel rockets, each 19 m long, fitted round a central 'core' rocket 28 m long and 3 m in diameter. All five of these rockets burn kerosene and liquid oxygen, develop a thrust of about 100 tons weight, and have four nozzles. On top of this assemblage is a final-stage rocket which goes into orbit with the satellite. Fig 4 shows a launcher of this type. The final-stage

Fig 4 *A Soviet launcher of the type most often used. The inset photograph shows the layout of the four extra rockets round the central core rocket. The final-stage rocket may be one of several designs, but in recent years has usually been a large cylinder, nearly 8 m long and about 2.5 m in diameter.*

Fig 5 *The solid-fuel US launcher Scout, before the successful launch of the Ariel 6 (UK6) satellite in 1979.*

'Cosmos rockets' are fine objects to observe, being visible to the naked eye at any distance less than about 800 km, and they are discussed further in Chapter 3. Strangely enough, the rockets of Sputniks 1, 2 and 3 in 1957-8 were brighter still, because no upper stage was used and the core rocket, 28 m long, went into orbit.

Not all the Soviet launchings call for the large launcher of Fig 4. Many of the smaller Cosmos satellites have a simpler-looking and lighter launcher, consisting of a first stage 20 m high and either 2.5 m or 1.65 m in diameter, topped by a second stage about 8 m long and of the same diameter as the first stage. These second-stage rockets go into orbit and are excellent objects for observing.

The prime example of a small solid-fuel rocket launcher is the American Scout, which has successfully launched more than a hundred small satellites, including the British spacecraft of the Ariel series. The Scout is 22 m long and 1 m in diameter, having a weight at launch of 21 tons. The final-stage rocket which accompanies the satellite into orbit is 1.5 m long and 0.5 m in diameter, with a mass of 24 kg after the fuel is burnt. Fig 5 shows the Scout launcher of the UK6 satellite, renamed Ariel 6 after its successful launch on 2 June 1979.

Another solid-fuel rocket launcher is the Japanese Mu, about 20 m high, with an initial weight of about 40 tons. Variants of the Mu have successfully launched six satellites.

Although the Saturn 5 launcher has left little for observers to track, there are several other US liquid-fuel launchers that have provided plenty of objects to observe. These are the Thor-Agena, Thor-Delta, Atlas-Agena and Titan-Agena launchers, which have together made more than five hundred successful launches. The Agena is one of the brightest of the US rockets in orbit and there are many to be seen, as described in Chapter 3.

Satellite launchers were also developed in the 1960s by France (the Diamant) and China. Of these the most important for the observer were the Chinese launches, which left in orbit large rockets rather similar to Cosmos rockets.

All the launchers so far mentioned date from the 1950s or 1960s. A large new launcher, which came into use in 1979, is the Ariane, developed by the European Space Agency. This is a three-stage liquid-fuel rocket which stands 47 m high, has a diameter of 3.8 m and a weight at launch of about 210 tons. The first and second stage motors use dimethyl hydrazine as fuel and nitrogen tetroxide as oxidant, while the third stage uses liquid hydrogen and liquid oxygen. The satellites launched will be near-equatorial and difficult to observe from middle latitudes.

Finally, there is the much-heralded Space Shuttle, the re-usable spacecraft with recoverable boosters, which is giving space flight a new image in the 1980s. The design differs from that of any previous launcher, because the winged Orbiter (discussed in Chapter 3) has its own rocket engines, which begin thrusting at launch, being fed with propellants from a huge external tank, 47 m long and 8.4 m in diameter, carrying 101 tons of liquid hydrogen and 603 tons of liquid oxygen. Two

Fig 6 *The Space Shuttle, showing the winged Orbiter, the huge external tank (47 m long) and the two recoverable boosters.*

solid rocket boosters are attached to the external tank (Fig 6): each is 46 m long and 3.7 m in diameter, weighs 586 tons (of which 504 tons is propellant) and has a maximum thrust initially of about 1300 tons weight. The boosters fall away after two minutes of flight and are recovered by parachute. The external tank is accelerated to just short of orbital speed and is then cast off to fall back into the atmosphere and burn up in a pre-arranged oceanic area. To an observer this seems a sad waste. If allowed to go into orbit, the external tank could have created fresh interest in space: it would have been the brightest of satellites, and its unpredictable burn-up could have been billed as a chance for gamblers as well as a free firework display.

The bright ellipse

These various launchers can give a satellite approximately the right speed for a circular orbit, but it is very difficult to achieve exactly the correct speed, so in practice nearly all orbits are ellipses rather than circles. The point on the ellipse where the satellite comes nearest to the Earth is called the *perigee*, and the point of maximum height the *apogee*, as shown in Fig 7. Most satellites are injected into orbit

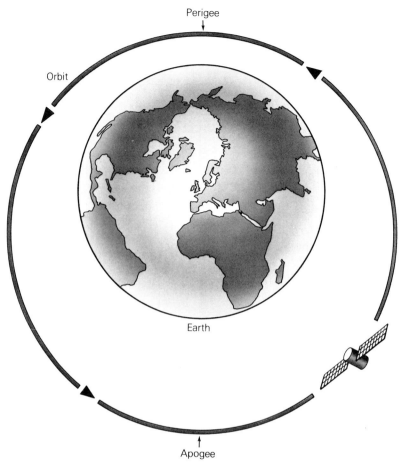

Fig 7 *Perigee and apogee points on an elliptic satellite orbit.*

at perigee, with apogee at the other side of the world, largely because this method of launching is simplest and the most economical in rocket fuel.

The size of a satellite's orbit is specified by its *average distance* from the Earth's centre, ie half the sum of the apogee and perigee distances. This average distance is usually given the symbol a, and the technical name for it is the *semi major axis*. If, for example, perigee is 7000 km from the Earth's centre and apogee is 9000 km, then $a = 8000$ km. For a circular orbit the apogee and perigee distances are equal, and the average distance a is merely the radius of the circle. For an elliptic orbit a can be thought of as the 'average radius' of the orbit.

The average radius of the orbit is very important, because it determines how long the satellite takes to go once round the Earth. This time is called the period of revolution, or, more simply, the *orbital period* of the satellite. When the satellite's average height is 200 km, which usually means that it will burn up in the atmosphere after a few days, the average radius *a* is 6570 km (taking the Earth's radius as 6370 km), and the orbital period is 88 minutes. If the average height is greater, so is the orbital period, in accordance with the law discovered by Kepler in the seventeenth century — that the square of the orbital period varies as the cube of the average radius. Thus, if the average height increases from 200 to 1000 km (*a* increases from 6570 to 7370 km), the orbital period increases from 88 to 105 minutes.

Fig 8 gives the orbital period when the average height is between 100 and 1800 km. The numbers on the curve indicate the average speed (equal to the speed in a circular orbit) in km per second. The satellites which are most interesting to observe usually have average heights of less than about 1500 km, so that their orbital periods lie between 88 and 116 minutes. Therefore *a near satellite takes between about 1½ and 2 hours to go round the Earth*. This is certainly a fact worth remembering; and if you want something more detailed, note that a period of 90 minutes implies an average height of about 300 km, 92 minutes implies 400 km, 94 minutes implies 500 km, and so on, up to 100 minutes (800 km).

The average radius of an orbit is enough to determine the orbital period, but we also want to specify the shape of the orbit: this is done by means of the *eccentricity*, defined as the apogee height minus the perigee height, divided by twice the average

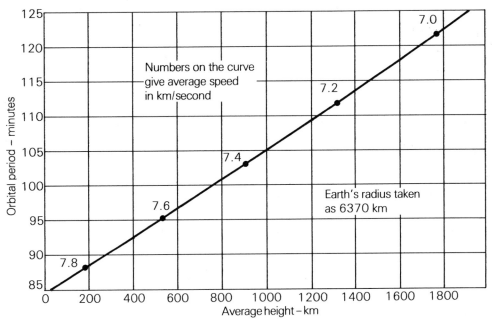

Fig 8 *Orbital period of a satellite versus its average height above the Earth.*

radius. Thus the more an orbit departs from a circle, the greater is its eccentricity. For the orbit with perigee and apogee points 7000 and 9000 km respectively from the Earth's centre, the eccentricity is $\frac{2000}{16000}$ = 0.125. The eccentricity is a useful quantity for specifying the shape because it always lies between 0 (for a circular orbit) and a little less than 1 (for an exceedingly elongated orbit). Most satellites of interest to observers have eccentricities between 0 and 0.2. If the eccentricity is 0.1, the apogee height exceeds the perigee height by about 1500 km for a near-Earth orbit.

We have seen how to specify the size and shape of an elliptic orbit by its average radius a and eccentricity e, but there is a third important orbital parameter — the inclination of the orbit to the equator. This is the angle at which the orbital plane cuts the plane of the equator and is usually denoted by the letter i. The inclination is important because it determines the maximum latitude reached by the satellite. Thus a satellite with an orbital inclination of 30° can always be found between latitude 30°N and 30°S, and if it is in a low orbit it can never be seen from latitude 50° — which is a pity, because there are several bright satellites in orbits with inclinations near 30°.

Life-span in orbit

A satellite moving in a nearly circular orbit at a height of 10000 km or more, if left undisturbed, will remain in orbit for thousands or even millions of years. If we destroy civilization by nuclear, chemical or biological warfare, such satellites would remain circling a devastated planet, relics of the advanced technology that led to our downfall. But perhaps we shall avoid this fate and establish a more rational society on Earth: if so, any unwanted satellites would probably be swept up to avoid pollution of the space environment. So the future of these long-lived satellites is arguable.

Satellites in lower orbits feel the effects of atmospheric drag, which slowly but inexorably reduces their energy. The density of the air falls off rapidly as you go up, until at a height of 200 km the density is ten thousand million (10^{10}) times less than at sea level. But the effect of air drag is still significant even at a height of 1000 km or more, and air drag governs the lifetimes of all objects in circular or moderately elliptic orbits if the perigee height is less than about 2000 km.

A satellite in an appreciably elliptic orbit suffers much more drag when near perigee than in any other part of its orbit, because the density of the air around decreases rapidly as the satellite climbs away from perigee. Air drag therefore acts almost as though concentrated entirely at perigee. Thus, if we return to our mountain peak on a non-rotating Earth, we can closely simulate the effect of air drag by holding up a paper hoop for the satellite to burst through each time it comes past. Bursting through the hoop slightly reduces its speed — the resistance of the paper representing the total effect of air drag. So the satellite is 'injected' into its next orbit at a

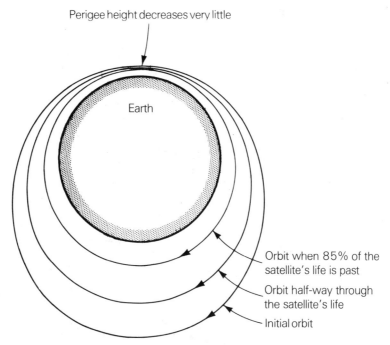

Perigee height decreases very little

Earth

Orbit when 85% of the satellite's life is past

Orbit half-way through the satellite's life

Initial orbit

Fig 9 *Contraction of a satellite orbit under the influence of air drag.*

slightly lower speed, and consequently does not fly off to quite such a great apogee distance on the other side of the Earth.

This is precisely the effect of air drag on an orbit. The perigee height remains almost constant but the apogee height steadily decreases. The orbit contracts and becomes more nearly circular, as shown in Fig 9. Since the average radius decreases, the orbital period also decreases, until finally, when the orbit is nearly circular at a height of about 200 km and the period is about 88 minutes, the satellite is ripe for its final fiery plunge through the atmosphere.

Experience at ground level tells us how drag affects satellites. The heavier the satellite, the less the effect of the air; the larger its area, the greater the influence of the air. Near the ground, a stone thrown is only slightly affected by aerodynamic forces, whereas a leaf is at their mercy. For a satellite this rule transforms into the fact that the lifetime of a satellite subject to air drag is directly proportional to its mass/area ratio m/s, where m is its mass and s its cross-sectional area. A small scientific satellite well stuffed with instruments might have $m = 100$ kg and $s = 1$ square metre, so that $m/s = 100$ kg/m^2, and this is also approximately the figure for Cosmos rockets, for which m is about 2000 kg and s is rather less than 20 m^2. So we may take $m/s = 100$ kg/m^2 as a standard value, though with the proviso that great variations from the standard are possible (balloon satellites being an obvious example).

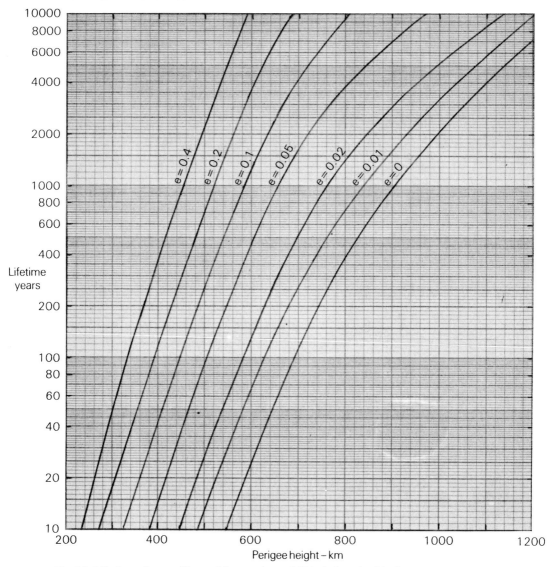

Fig 10 *Lifetimes for satellites with mass/area 100 kg/m² and orbital eccentricity e.*

As we shall see in Chapter 10, the density of the upper atmosphere is very variable, and at heights near 500 km may be ten or twenty times greater at the maximum of the eleven-year sunspot cycle than it is at the minimum. So the prediction of a satellite's lifetime can never be exact. Still, a rough guide is better than none, and Fig 10 shows the approximate lifetime, with average solar activity, of a 'standard' satellite with mass/area of 100 kg/m². The lifetime is greatly dependent on both perigee height and apogee height: Fig 10 gives a life of 25 years for a circular orbit at a height of 600 km; but increasing the height to 1000 km (still circular) extends the lifetime to 2000 years. If the perigee height is 600 km and the apogee height 1000 km, the eccentricity is about 0.03 and the lifetime about 250 years.

If the mass/area differs from 100 kg/m², the lifetimes in Fig 10 should be scaled accordingly. For example, a satellite with mass/area 50 kg/m² will have half the lifetime shown in Fig 10. If the lifetime obtained from Fig 10 is less than ten years, the assumption of average solar activity may be incorrect, depending on the date of launch of the satellite. The last maximum of solar activity occurred in 1981, but forecasts of the dates of future maxima and minima are notoriously unreliable. The next minimum is expected about 1986, followed by a maximum about 1990 and another minimum about 1996.

Air drag does not always control a satellite's lifetime, however. If the orbital eccentricity is greater than about 0.6, so that the apogee height exceeds 20000 km, the gravitational pull of the Sun and Moon can cause great changes in the perigee height and may force perigee down to heights below 100 km, where air drag rapidly destroys the orbit. It is not possible to give any general rules for the effect of the Sun and Moon: it all depends on the geometry of the orbit relative to the Sun and Moon, and how this geometry changes as the years go by. Near-equatorial orbits are affected much less strongly than near-polar orbits, for which the changes are often very large. The Russian Molniya satellites, for example, of which more than a hundred have been launched, have an inclination of 63°, an initial perigee height near 400 km, and an eccentricity of 0.75: the Sun and Moon usually pull the perigee up to about 2000 or 3000 km within a few years, only to send it plunging back again, usually giving lifetimes between ten and twenty years. When the lifetime is controlled by the lunar and solar gravitational perturbations, the mass/area ratio of the satellite is of very little importance.

A final paragraph on balloon satellites is needed, because their lifetimes are usually controlled by another mechanism, the pressure of sunlight. Though extremely weak, solar radiation pressure is stronger than air drag at heights above about 600 km, and can alter the perigee height of a balloon satellite by 100 km or more over a period of years. Since its mass/area is very small, perhaps 1 kg/m², a balloon satellite has a much shorter lifetime than normal satellites at comparable heights. For example, Fig 10 shows that a balloon satellite with perigee height 800 km and eccentricity 0.05 would have a lifetime of 40 years if the mass/area was 1 kg/m², rather than 4000 years if $m/s = 100$ kg/m². But this lifetime estimate must itself be modified to allow for the changes in perigee height caused by solar radiation pressure.

The fiery finale

A satellite's life always ends with a spectacular incandescent entry into the lower atmosphere and, unless specially designed to withstand the heat, the satellite breaks up and burns up after providing a fine fireworks display for anyone who may be watching below.

Because the air density in the upper atmosphere varies by about 10% from day to day in an irregular and largely unpredictable manner, the descent date of a satellite — or 'decay' as it is usually called — can only be predicted with an accuracy of about 10% of the remaining life. So a prediction made ten days before decay is likely to be in error by about one day, and even a prediction made twenty-four hours before decay be in error by 2 or 3 hours, in which time the satellite goes twice round the world. Slightly better predictions may be possible if the satellite can be observed on its last day in orbit, and the observations can be passed instantaneously to the predictor; but the cost of these operations is not usually regarded as justifiable. Even when all possible care is taken, it is only the track over the Earth during the last revolution that can be predicted, not the exact geographical location of the burn-up. The uncertainty arises because the satellite may rotate in an unexpected way on meeting the denser air, and alter its drag or generate aerodynamic lift which will deflect it from the calculated path. Because of this uncertainty, everyone, observer or no, has a chance of seeing a satellite burn up (though those with cloudless skies have the best chance, of course). People who have seen a satellite burn up all agree that it is a spectacular sight.

Fig 11 *Drawing of the descent of Cosmos 253 rocket over England on 20 November 1968.*

Fig 12 *Photograph of the descent of Skylab 1 over Australia on 11 July 1979.*

The first satellite descent to be observed was that of Sputnik 2 over the Caribbean Sea on 14 April 1958. The satellite, 32 m long and 3 m in diameter, was visible for 5 minutes on a track running from New York to near the mouth of the Amazon. It developed a long trail of sparks and took on nearly every colour of the rainbow, to judge from the numerous eye-witness reports.

The most widely observed burn-up over Britain so far has been that of the rocket of Cosmos 253 on 20 November 1968. The rocket was 7.5 m long and 2.6 m in diameter, and it came down on a line running from Manchester to Southend and Dover. Fig 11 shows a sketch of its appearance: the colour was primarily white or whitish-yellow, with touches of orange, red, green and blue.

The most publicized burn-up was that of the 75-ton Skylab 1 over the Indian Ocean and Australia on 11 July 1979. It was an awesome sight, and Fig 12 is a photograph which gives an idea of its appearance.

When a two-ton satellite enters the atmosphere, usually at a speed close to 8 km per second, it possesses prodigious energy of motion, roughly equal to that of 50000 cars all travelling at 100 km per hour. This is the energy imparted by the launching rocket (minus a small proportion already whittled away by air drag). Nearly all that energy is converted to heat and light as the satellite plunges into the thicker air. It usually begins to glow at a height of about 90 km and continues its firework display down to a height of about 30 km. By then it has lost most of its initial velocity, and any remaining fragments begin to fall at a steep angle.

A small satellite usually burns up completely, but some fragments survive when the satellite is large, like Sputnik 4, which came down at Manitowoc, beside Lake Michigan, on 5 September 1962. The largest piece of Sputnik 4 found on the ground was a chunk of metal weighing 10 kg, but there were also many small charred fragments. Fig 13 shows a selection of these, recovered from the roof of the Lutheran Church: they look rather as if fire and brimstone had dropped from heaven; and that is more or less what did happen.

With a very large satellite like Skylab 1, which was 25 m long and 6.6 m in diameter, much larger pieces survive, because objects inside the main shell of the spacecraft are protected from the heat until the outer shell breaks up; so they begin their own independent descent at a relatively low speed, with a good chance of survival. Fig 14 shows one of the larger fragments from Skylab 1 found in south-west Australia.

No one has yet been injured by a falling satellite, and the chance of such injury

Fig 13 *Small fragments from Sputnik 4 found on the roof of the Lutheran Church at Manitowoc, Wisconsin, where the satellite came down on 5 September 1962.*

```
0                        1                        2
└────────────────────────┴────────────────────────┘
           Scale    in    inches
```

is extremely remote. Far more dangerous are aircraft, of which at least a hundred per week fall out of the sky, worldwide, as compared with about two satellites per week; and the aircraft are much more likely to be over populated areas. With an expected casualty rate of about one person per hundred years, as compared with several hundred people *per day* on the world's roads, space travel deserves to be called the safest form of transport for bystanders, as well as enjoying zero fuel consumption once in orbit.

Fig 14 *A large piece of Skylab 1 found in Australia after its descent on 11 July 1979.*

3
Traffic
of the Sky

And drives o'er Night's blue arch his glittering car.

Erasmus Darwin, *The Botanic Garden* (1792)

The first 'glittering car' to drive along the orbital race track above the dark arch of the Earth's shadow — the Russian Sputnik 1 — was launched on 4 October 1957. Since then, the traffic on the space track has become gradually more congested, until in the early 1980s the merry-go-round of orbital traffic consists of about 5000 vehicles, ranging in size and mass from a double-decker bus down to a child's roller-skate. From this incongruously wide variety of objects, satellite observers can pick and choose. Some observers like to look at random, on impulse, but an informed choice is usually more productive, so I shall attempt a quick survey of the traffic in orbit, after first asking how and why the traffic built up, and seeing just how brightly the satellites glitter.

To show how the traffic has built up over the years, Fig 15 gives the number of launchings of satellites and space vehicles each year from 1957 to 1981, divided into three categories: (1) USSR, (2) USA and (3) 'others', if any. The 'others' include launches by other countries, and also launches by the USA or USSR on behalf of other countries or international bodies, such as the Intelsat Corporation. Fig 15 shows that the USA had more launches than the USSR in each year between 1958 and 1966, but the situation reversed in 1967, and the preponderance of launches by the USSR beginning in 1967 continued throughout the 1970s. In the late 1970s the proportions were approximately: 75% USSR; 15% USA; and 10% 'others'. When the Space Shuttle is in full operation, the proportion of US launches may increase again.

After the USSR and USA, the next country to launch a satellite with a home-made

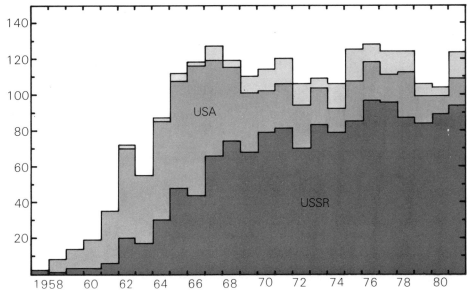

Fig 15 *Yearly numbers of satellite launchers, 1957 to 1981, divided into USSR, USA and 'others'.*

rocket was France (1965), followed five years later in quick succession by Japan (1970), China (1970) and the UK (1971). Then there was a long interval until India joined the league in 1980. (The European Space Agency's first Ariane was launched in 1979, but that scarcely counts as a national launch.)

Fig 15 also shows that the number of launches per year increased rapidly between 1958 and 1966, when the annual total reached 118; but then, rather surprisingly in such a changing world, the figure remained nearly constant for fourteen years, always being between 105 and 128, with an average of 116. The total of launches to the end of 1981 was 2268.

The number of objects in orbit resulting from all these launches has been rather unpredictable because a fairly small number of satellites have exploded in orbit — sometimes by design, sometimes by accident — giving rise to perhaps 10 or perhaps 100 fragments. From the 2335 launches up to 1 July 1982, a total of 13300 pieces were tracked at one time or another. During the 1970s about 43% of the total number of objects were still in orbit, the rest having decayed. The strong solar activity in 1979-81 increased the density of the upper atmosphere and hastened the decay of all satellites affected by drag; consequently the percentage in orbit was reduced, and stood at 35% in 1982.

Launching a satellite is an expensive exercise, and if we (optimistically) assume that human affairs retain some relics of rationality, we may reasonably ask why all these satellites were launched. The difficult task of assigning the launches to particular categories has been tackled by Dr Charles Sheldon and I shall try to summarize his findings.

He divided the launches into two main groups, military and civil. About 60% of the launchings have been primarily military, the proportion being greater for the USSR than for the USA. In this military group, more than half have been photographic-reconnaissance satellites, about 10% have been for communications, rather less than 10% for navigation, and the remainder for a variety of purposes, such as early warning of missile attack, ocean surveillance, electronic 'listening in', and tests of satellite interception.

If we now turn to the 40% of launchings that have been primarily civil, the subdivision is as follows. About 40% of them have had scientific research as their aim: most of these satellites were designed to examine the Earth and its environment; other targets for research have been the planets, the Sun, the Moon or the stars. Nearly 20% of the civil group have been communications satellites, more than 10% weather satellites, and rather less than 10% manned satellites. Among the others have been many development satellites, for testing new instruments or engineering techniques, and a few satellites for the mapping of Earth resources.

'Have all the 2000-plus launches been really necessary?' is another question that may seem reasonable, but is scarcely worth asking. A project once approved develops a momentum of its own in countries with strong bureaucracies or in military organizations with a strong command structure, where an 'on-going programme' is the rule, rather than the 'start-stop' syndrome which plagues democratic bodies. The former is wasteful, because too many satellites are launched, but efficient in having a production-line run. The latter is wasteful because so much effort is expended in gaining approval for a single launch, when a planned series of launches would be far more efficient.

Pet names and proper names

Each launching organization likes to give each of its satellites a 'pet name' to be used by those working with the data sent back by the satellite. Many of these names are familiar — Apollo, Ariel, Cosmos, Skylab and so on — but some are unintelligible acronyms, like SPADES (*s*olar *p*erturbation of *a*tmospheric *d*ensity *e*xperiments *s*atellite) or SESP (*s*pace *e*xperiments *s*upport *p*rogram). Sometimes there is confusion of pet names: for example, the satellites Geos 1 and 2 (*ge*odetic *s*atellite) were launched by the USA in 1965 and 1968 and are still being used; this did not prevent the European Space Agency (ESA) from launching two satellites for studying charged particles in space, also called Geos 1 and 2 (*ge*ostationary *s*atellite) in 1977 and 1978. Ironically, ESA-Geos 1 was (unintentionally) sent into an orbit very far from geostationary, so the name was inappropriate. Even more ironically, the duplication has ruined ESA's own computerized monthly bibliography, which has regularly included all papers on ESA-Geos under the heading 'geodetic satellites'.

With the aim of avoiding such confusion, the international Committee on Space Research (known as COSPAR) has given all satellites and fragments an international designation based on the year of launch. Thus the British satellite Prospero is designated 1971-93A, because it was launch 93 of the year 1971. Usually the letter 'A' is given to the instrumented spacecraft, 'B' is given to the rocket, and C, D, E, and so on, to fragments (the letters I and O are omitted). Thus 1971-93B is the rocket that accompanied Prospero into orbit, and 1971-93C is a fragment — an aerial that was knocked off during injection into orbit. If several spacecraft are sent into orbit in one launch, they are given the letters A, B, C, etc: in the many 8-satellite launches by the USSR, the satellites are A-H and the rocket is J (remembering that I is omitted). When there are more than twenty-four pieces from one launch, as can happen after explosions, the sequence continues after Z with AA, AB, AC . . . AZ, and then BA, BB, BC . . . BZ and so on. The greatest number of fragments so far recorded from one launch is 462, resulting from the explosion of the satellite 1965-82A. (Before 1963, the designations were in terms of Greek letters rather than numbers, the first launch of 1960 being 1960 alpha, the second 1960 beta, etc.)

Besides reducing confusion, the international designation is useful because it gives not only the year but also the approximate month of launch. Since 1967 there have been about 120 launches per year, or 10 per month, so you might guess that Prospero (1971-93A) would have been launched in the tenth month of 1971: it was in fact launched on 28 October 1971. Obviously, there may be an error of one or two months because of the variations in the annual numbers of launches.

Relying on this rough rule, I shall often give the international designation instead of the date of launch, thus supplying the correct designation of the satellite and an approximate indication of its launch date.

How bright?

Before we can start choosing satellites for observation, we need to have some idea how bright they are: it is embarrassing to start a campaign of observation on a satellite that proves to be virtually invisible.

The brightness depends mainly on distance, size, shape, surface finish and 'phase angle' relative to the Sun. It is easiest to begin with spherical satellites: the largest of these has been the Echo 2 balloon (1964-04A), which had a diameter of 41 m and a mass of 256 kg; one of the smallest has been Musketball (1971-67D), with a diameter of 0.3 m and a mass of 61 kg, a football-sized lump of solid metal.

If you observe a spherical satellite in the east when the Sun is just below the horizon in the west, nearly all the surface of the satellite facing you is illuminated, like a full Moon (the largest spherical satellite of all). If you view the satellite in the west when the Sun is also in the west, only a crescent is illuminated, like a new Moon. Obviously, a satellite will be much brighter as a 'full Moon' than as a crescent, and to provide a standard we choose the situation when the satellite is half illuminated,

27

like the Moon at 'first quarter'; that is, when the 'phase angle', the angle 'Sun-satellite-observer', is a right angle. If the satellite is fully illuminated it will be twice as bright as the standard; but if it is a crescent it may be very much fainter than the standard. You have the best chance of seeing a faint satellite if you look for it at a point on its track where the phase angle is a right angle or greater, and this is a factor that all observers keep in mind.

The influence of 'surface finish' on brightness is much less amenable to logic. Needless to say, a satellite painted black is fainter than one painted white or having a shiny metallic surface. But as time goes on, a white or shiny surface becomes tarnished, while a black one may become brighter, if the paint flakes off. Observers cannot go up into orbit to examine the surface, and they are content to establish the brightness of each satellite by observation, noting any changes as the years pass.

Since satellites look like moving stars, their brightness is specified in the same way as that of stars. Most people know that 'a star of the first magnitude' is a bright one, and the basis for the 'magnitude' system of star brightness is that magnitude 1 should represent the average brightness of the twenty or so brightest stars; while, at the other extreme, magnitude 6 applies to the faintest stars visible to the naked eye on a clear dark night. The first-magnitude stars are about 100 times brighter than sixth-magnitude stars, so the scale of magnitudes has to be such that 5th magnitude is $2\frac{1}{2}$ times brighter than 6th, while 4th is $2\frac{1}{2}$ times brighter than 5th, and so on, as shown below.

Magnitude of star	6	5	4	3	2	1	0	-1
Brightness (6th mag = 1)	1	2.5	6.3	16	40	100	250	630

For the brightest stars, planets and satellites, negative magnitudes are needed. The star Sirius, the planet Mars, the Russian space station Salyut and the Space Shuttle Orbiter are often of magnitude -1, that is $2\frac{1}{2}$ times brighter than a star of magnitude 0, such as Vega, which is itself $2\frac{1}{2}$ times brighter than a star of magnitude 1. For intermediate brightness a decimal point is introduced: Capella is of magnitude 0.2, the Pole Star is of magnitude 2.1, and the seven stars of the Plough, beginning at the handle, are of magnitudes 1.9, 2.4, 1.7, 3.4, 2.5, 2.4 and 2.0.

In practice it is useful to specify a 'standard magnitude' for every satellite, defined as its magnitude when half illuminated at a distance of 1000 km, which is about the average distance for a satellite observation. A spherical satellite of diameter of 1 m usually has a standard magnitude of about 8; and a spherical satellite of diameter 10 m, having 100 times the area, would be 100 times brighter, that is magnitude 3. Similarly, the effect of distance can be allowed for: bringing a satellite to $\frac{1}{10}$ of the distance makes it 100 times, or 5 magnitudes, brighter. It is worth noting that if either diameter or distance changes by a factor of 2, the magnitude changes by $1\frac{1}{2}$; if the factor is 4, the change in magnitude is 3; if the factor is 10, the change in magnitude is 5. Also, of course, the two effects cancel each other, so that a sphere 3 m in diameter

300 km away has the same brightness as a sphere 5 m in diameter 500 km away.

These rules are used to construct Fig 16, which shows the brightness of half-illuminated spherical satellites of various diameters at various distances, on the assumption of average reflectivity. A satellite painted black would be less bright, while a shiny one would be somewhat brighter than the diagram indicates. Fig 16 applies for satellites that are high in the sky — at elevations greater than 30°. Near the horizon a satellite is fainter because its light has to travel a long way through the (usually) polluted lower atmosphere. Unless the air is exceptionally clear, a satellite

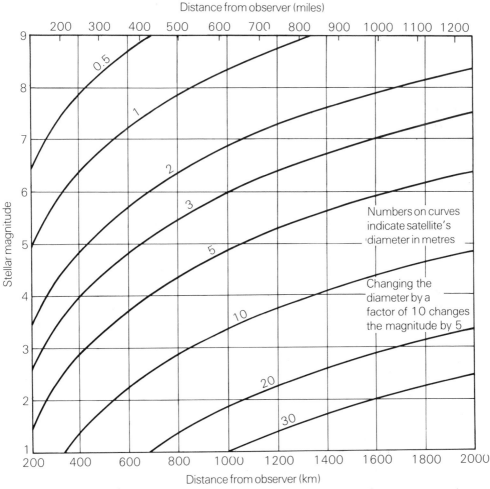

Fig 16 *Brightness (stellar magnitude) of spherical satellites of various diameters at varying distances. Those of magnitude 1–5 are visible to the naked eye on a clear moonless night, and those of magnitude 8 or brighter can be seen with 7 × 50 binoculars.*

at a particular distance appears about 1 magnitude fainter at 15° elevation than when overhead.

The majority of satellites observed are not spherical, but cylindrical, and their brightness can be approximately estimated by calculating the maximum area of the cylinder, as seen from the side, ie length × diameter, and then taking the square root, to give the diameter of an equivalent sphere. For example, a rocket 8 m long and 2 m in diameter has a side area of 16 m² and would have about the same brightness as a sphere of diameter 4 m.

This method of estimating the brightness of a cylindrical satellite makes some allowance for the observer not seeing it exactly sideways-on. But sometimes you may see it end-on, and it then appears much fainter. In practice, the majority of cylindrical satellites rotate: their motion is usually somewhere between spinning like a propeller and tumbling end-over-end; the axis of rotation changes its direction in space only slowly from day to day, and is usually unrelated to the direction of travel. Cylindrical satellites sometimes rotate rapidly, eg once per second, but the rate is usually much slower, eg twice per minute. As their aspect changes, so does their brightness, and they generally exhibit a regular fluctuation in brightness: observers are asked to record the maximum and minimum magnitudes, and also the 'flash period', so called because the peaks in brightness are like flashes of light if the rotation is rapid.

Balloons — inflated, in shreds or in absentia

Any survey of satellites for observing should begin with balloon satellites. Throughout the 1960s and 1970s they proved themselves the most reliably bright for observers, and the large Echo 2 balloon, 41 m in diameter, in orbit from 1964 to 1969, was probably seen by more people than any other manufactured object in the history of the world. Balloons are also the cheapest satellites to manufacture, and provide free publicity to the launching country for many years. So it is logical to expect that there would be plenty of balloons to observe. But logic is rare in our world — bureaucracy 'moves in a mysterious way its blunders to perform' — and I have to admit that, as I write, not a single balloon is to be seen.

But I still cling to my outmoded faith in logic, despite all the evidence to the contrary, so I expect to see more balloon satellites. If I am right, they may be similar to two balloons of the 1960s which both decayed in 1981, namely Explorer 19 (1963-53A), shown in Fig 17, and Explorer 39 (1968-66A). Both were spheres 3.6 m in diameter, with masses of 7 kg and 9 kg respectively. Because of their low mass/area ratio — about 1 kg/m² — the orbits of these balloon satellites were very sensitive to air drag and they were launched in the hope that analysis of their orbits would advance knowledge of the upper atmosphere at heights of 600-800 km. This hope was fully realized. Balloon satellites are also of great interest to orbital analysts because the effects of solar radiation pressure, and Earth-reflected radiation pressure, are

Fig 17 *The Explorer 19 balloon satellite, which was 12 feet (3.6 m) in diameter. Launched in 1963, it decayed in May 1981; but future balloon satellites may be of similar design.*

difficult to calculate, and still give the theorists headaches. They need real-life examples to work on.

Initially, Explorer 19 had a perigee height near 600 km and an apogee height of 2500 km, but solar radiation pressure caused the perigee height to oscillate over the years between a minimum of 600 km and a maximum of 700 km. Explorer 39 was in a rather similar orbit initially. Both balloons were visible at magnitude 5 at a distance of 1000 km, and were excellent objects for observing.

Though these two balloons retained their spherical shape until decay, there is a larger balloon called Pageos (1966-56A), which has disintegrated and is in shreds. Pageos was an aluminium-coated plastic balloon, 30 m in diameter with a mass of 55 kg, launched into a circular orbit at a height of 4200 km, inclined at 87° to the equator. Pageos was a *passive geo*detic *s*atellite, intended to provide a point of light in the sky for use with cameras of high accuracy, to link the mapping networks of countries widely separated, with an accuracy of about 10 m. Pageos achieved its purpose for nearly ten years, and made a fine contribution to geodetic science; and then, as if unable to stand the strain any longer, it broke up into 28 pieces on 12 July 1975. The largest piece was still almost as bright as the original satellite, but on 20

January 1976, a further 44 fragments appeared. Just after the second break-up, a line of fragments could be seen trailing one behind the other like shining beads on a string, but after a few days they became separated. Now all that remains are numerous shreds of plastic circulating in orbit. The fragments are still fairly bright at times, and some observers find it fascinating to try to keep track of them. There were various speculations about the cause of the first break-up, but the subsequent disintegration suggests that the plastic mylar of which Pageos was constructed had after ten years in space more or less fallen apart — the plastic, if not bio-degradable, was certainly space-degradable.

Cosmos rockets

After these flimsy balloons we have something much more substantial, of solid metal. Most of the Russian satellites are given the name 'Cosmos' and, although the name covers a multitude of different purposes, from military reconnaissance to pure scientific research, there is a logic in it from the observer's viewpoint, because very similar second-stage rockets go into orbit with nearly all Cosmos satellites, see Fig 18. There are some variations among the rockets, depending on the launcher used, and sometimes one rocket will launch several satellites: but all these rockets have had diameters between 1.5 m and 2.6 m, and lengths between 6 m and 8 m, so it is difficult to distinguish between them observationally. There have been more than a thousand Cosmos rockets in orbit and they have provided a 'staple diet' for observers for many years. More than half of the thousand have been in low orbits with lifetimes of two weeks or less, but the rest of these Cosmos rockets travel in widely varied orbits at inclinations between 48° and 83°. About two hundred are in long-lived orbits, where they will remain a feature of the night sky until they are swept away by an unfriendly power or a Soviet government keen on clearing up space.

Cosmos rockets are bright enough to be seen with the naked eye whenever they come within about 1000 km of an observer — their standard magnitude being about 4. Many of them move in stable orbits at heights of 800-1000 km and are easy to observe with the aid of binoculars because they can be accurately predicted. Those with low perigee and high apogee are much more useful, however, because the changes in their orbits caused by air drag reveal the density of the upper atmosphere near perigee and the atmospheric winds (see Chapter 10). The rockets in these eccentric orbits are also more of a challenge to the observer, not only because the variations in air density may upset the predictions, but also because the satellite may rush past very rapidly at perigee or be rather faint at apogee.

Many of the Cosmos rockets rotate quite rapidly, and consequently their brightness varies regularly. Soon after launch, the brightness may vary once per second, or even faster, but the rotation becomes slower as time goes on. If a number of observers in different places record the maximum and minimum brightness, it is possible to determine the direction of the spin axis in space. So the drift of the axis

Fig 18 *Most Soviet launchers of Cosmos and Soyuz satellites have a cylindrical final-stage rocket that goes into orbit, the middle section of the rocket shown here. There are various designs, but the length is usually between 7 and 8 m and the diameter between 1.5 and 2.5 m.*

over the months can be followed, and compared with that expected from theoretical studies. Timing of the 'flash period', the interval between successive maxima of brightness, is also of interest in showing how the spin dies away over the months. It can tax the powers of the most skilful observer to make three accurate positional observations of a low-perigee Cosmos rocket, to measure the flash period and to estimate the maximum and minimum brightness in two different parts of the sky — all within a minute or two.

Not all the Cosmos rockets rotate rapidly, and some shine quite steadily. Usually they are seen more or less 'side-on', but sometimes you may see them end-on, and much fainter. So although you can rely on seeing a rapidly spinning Cosmos rocket, because every few seconds it shines at magnitude 4 or brighter (if within 1000 km), a slow spinner may elude you by remaining faint during the whole of the time taken to cross the field of view of your binoculars.

All in all, the Cosmos rockets provide quite a range of challenges — enough to keep an observer occupied even if there were no other satellites in the sky at all.

Cosmos satellites

After the Cosmos rockets it is only logical to look at the spacecraft they have launched — 1400 of them by August 1982, of a wide variety of shapes and sizes.

The most numerous have been the 5-ton photographic reconnaissance satellites, which consist of a sphere of 2.4 m diameter (similar to the Vostok spacecraft which carried the first man in orbit, Yuri Gagarin) and a cylindrical section of the same diameter, the total length being about 6 m. The spherical 'Vostok' section can survive re-entry into the atmosphere, and is recovered after a week or two in orbit — together with its film of the Earth beneath. During the 1970s, more than three hundred of these satellites were launched, one every ten days or so, on average. In the 1960s their favourite time-of-stay in orbit used to be 8 days, but times-of-stay between 12 and 14 days have become more usual since 1971. The favourite Russian orbital inclination of 65° was used for 44% of the launches in the 1960s and 1970s; 16% were at 73°, 14% at 63°, and 11% at 81°, with inclinations of 70–74° and 81–83° becoming dominant in the early 1980s.

These reconnaissance satellites move in low orbits with perigee height near 200 km and apogee height near 300 km (on average), and nearly always have orbital periods between 89 and 91 minutes. They are easily visible to the naked eye, being of magnitude about 2, and they cross the sky quite rapidly, taking about 3 minutes from horizon to horizon. Though easy to see, these satellites are not easy to predict, partly because they return to Earth before the prediction centre has time to issue regular weekly predictions, and partly because many of the reconnaissance satellites have small rocket motors for altering their orbits.

Another type of Cosmos satellite is the much smaller spacecraft used for scientific research. There have been more than a hundred of these, about 1.8 m long and 1.2 m in diameter, with a mass of about 400 kg. Their orbits are varied, and often useful for upper-atmosphere studies. But the rocket that enters orbit with the spacecraft is brighter and is usually given preference for observing. Consequently the small Cosmos satellites have had rather a raw deal from observers, though they are quite easily observed, usually being of magnitude 6 or 7 at a distance of 1000 km.

Not all scientific experiments can be fitted into these small satellites, so there are many Cosmos satellites, both scientific and military, which are almost as large and bright as their rockets, or have remained attached to their rockets so that they are even brighter, like Cosmos 893 (1977-11A). The most useful of these for research purposes, as with the rockets, are those with low perigee and high apogee. Two satellites at 51° inclination launched in the early 1970s — Cosmos 379 (1970-99A) and Cosmos 398 (1971-16A) — have served this purpose particularly well. Their perigee heights are near 200 km and their apogee heights were initially above 11000 km. They offer quite a challenge to observers because of the variations as their height slowly changes from perigee to apogee. At perigee they cross the sky rapidly, travelling at up to 2° per second, and often go into eclipse while still high in the sky: making two or three good observations so quickly is a real test of skill. At apogee, on the other hand, the satellites are slow-moving and faint, so the challenge then is to make exceptionally accurate observations by utilizing stars as faint as the satellite, of magnitude 8 or 9.

With so many Cosmos satellites in orbit, some of them will happen to enter orbits which are particularly useful for geophysical studies. For example, in the early 1970s many Cosmos satellites in circular orbits at 74° inclination repeated their track over the Earth daily after 15 revolutions. As a result they were very valuable for studying the Earth's gravitational field, as we shall see in Chapter 9. For these and other reasons, many Cosmos satellites have become 'targets of opportunity' for visual observers — satellites intensively observed for a few months or perhaps a year or two, for a specific scientific project.

Agena rockets

The nearest American equivalent of the Russian Cosmos rocket is the Agena rocket, which was used as an upper stage in hundreds of launchings during the 1960s and 1970s, when the Thor-Agena, the Atlas-Agena and the Titan-Agena were the chief launchers for the heavier US payloads. Many of these Agenas have already decayed, but more than a hundred are still in orbit.

An empty Agena rocket (Fig 19) is 1.6 m in diameter and 7 m long with a mass of about 700 kg. The fuel is dimethyl hydrazine and the oxidant is nitrogen tetroxide. Often the rocket separates from the (one or more) spacecraft that it launches, but sometimes the payload remains attached to the rocket in a nose cone, which increases the length, typically to about 8 m. On a near transit, at a distance of about 250 km, an Agena is usually of magnitude 2, about as bright as the Pole Star. But the majority

Fig 19 *The US Agena rocket, of which over 300 have gone into orbit, and more than 100 remain there. The diameter is 1.5 m and the basic length 6 m, but this satellite – the Target Agena 8 used in the first docking of two satellites in 1966 – has extra equipment fitted at the front. This photograph was taken in orbit, at a distance of 14 m.*

of the Agenas are in long-lived orbits at much greater heights, most often in near-polar orbits at heights near 1000 km. So Agenas are most often seen travelling roughly north to south (or south to north) at a magnitude of about 5.

A number of Agena rockets are among the 'targets of opportunity' observed for specific geophysical researches. One of these is 1964-01A, which is in a circular orbit at a height of 900 km, inclined at 70° to the equator. Another Agena, which was a target of a different kind, is shown in Fig 19. This is the Target Agena 8 used for the first docking of two satellites in 1966: Gemini 8, with Neil Armstrong and David Scott aboard, docked with Target Agena 8, but the linked spacecraft began tumbling soon afterwards and had to separate. The Agena in Fig 19, which was photographed at a distance of only 15 m, had special equipment for the docking instead of a nose cone.

Seasat

An Agena-rocket-plus, with the 'plus' larger than the Agena, is Seasat, which is the rather weird-looking spacecraft shown in Fig 20. Seasat, 1978-64A, carried a radar altimeter which measured its height above the ocean surface accurate to about 10 cm. Thus the profile of the ocean surface could be (and was) obtained more accurately

Fig 20 *Seasat, which was launched in June 1978 and for three months successfully measured its height above the sea surface accurate to about 10 cm, using a radar altimeter. The satellite is 21 m long and 1.5 m in diameter and is a bright object in the night sky.*

than ever before: Seasat gave scientists their first world-wide pictures of ocean currents, wave heights and wind speeds. The instruments on Seasat failed in October 1978, and since then it has been circulating in orbit undisturbed. It is a fine object for observing, because it is 21 m long and 1.5 m in diameter and is usually visible to the naked eye as it passes over at a height of 800 km at magnitude 3.

Seasat is in a near-polar orbit, but at an inclination greater than 90°, actually 108°. So, although its motion seems to be approximately south to north (or north to south), Seasat has a slight westward movement, instead of travelling eastward like the great majority of satellites. Another curious feature of Seasat is that its track over the Earth nearly repeats every 3 days after 43 revolutions.

Ariel, Prospero and Miranda

After looking at the hundreds of Russian and American rockets, let us now have a scene-change and bring on the small troop of British satellites, all named after characters in Shakespeare's magical late play *The Tempest*.

Ariel 1 was launched in April 1962, and was the first 'international' launch — by one country for another — being launched by a US Thor Delta rocket. It was the first of a series of six Ariel satellites, each devoted to some different aspect of geophysical or astronomical research, and each weighing approximately 100 kg. The last, Ariel 6, was launched in 1979 and is the only one still in orbit.

Prospero enjoys the distinction of being the only satellite sent into orbit by a British launcher; and Britain is unique in being the only country to launch only one satellite. Both these distinctions are likely to remain for many years to come. Launched by Black Arrow on 28 October 1971, the 66-kg Prospero is a pumpkin-like satellite 1.1 m in diameter and 0.7 m high, shown in Fig 21 with the final-stage Waxwing rocket attached. Shakespeare's Prospero had a magic wand, or staff; the satellite Prospero had four, in the form of four aerials to transmit its data, as if by magic, to the Earth. At the end of Shakespeare's play Prospero says, 'I'll break my staff', and abjures his magic; the Waxwing rocket seemed set on proving Shakespeare right, for, after injecting Prospero into orbit and separating, the rocket flew on and collided with Prospero, breaking one of its four aerials, though the other three remained in action. The resulting asymmetry in the radio signals provided an excellent though unintended method for measuring the spin rate. Strangely enough, the broken aerial will probably stay in space longer than any of the six Ariels.

Prospero was designed to measure the impact of micrometeorites, and the performance of solar cells and paints in space. The design life of one year was handsomely exceeded: for many years Prospero has been switched on for its birthday, and has obediently transmitted data to its home station at Lasham in Hampshire. Both Prospero and its rocket are rather faint objects, being of magnitude 7 when observed high in the sky near perigee (550 km height) and magnitude 9 near apogee (1500 km). Their orbital inclination is 82° and their lifetimes are expected to exceed a hundred years.

Fig 21 *The Prospero satellite (black) launched by the British Black Arrow rocket on 28 October 1971, with its final-stage Waxwing rocket (white) attached. The satellite and rocket, which separated on enter-ing orbit, should continue circling the Earth for about a hundred years.*

Miranda (1974-13A), the successor to Prospero, is a 93-kg technological satellite intended to test a new attitude control system. Since Black Arrow no longer existed in 1974, Miranda was launched by a US Scout rocket, into an orbit inclined at 98° to the equator. With perigee height 700 km and apogee 900 km, it should survive in orbit for more than a century.

The life-cycle of British civil satellites is now apparently complete, and, looking back, the story is like a five-act tragedy, a saga of diminishing expectations. Act I came in the early 1960s with plans for a UK launcher called Black Prince, consisting of the Blue Streak rocket as first stage and Black Knight as second stage, which could have lauched a 1000-kg satellite. In Act II, these plans were scaled down into Black Arrow, in which the first stage was vitually two Black Knights, and the second stage half a Black Knight: a satellite of about 100 kg could be (and was) launched. In Act III, Black Arrow was cancelled, chiefly because the committees of University researchers preferred the free launches offered by the USA, naïvely failing to realize that the free launch would become an expensive one if the rival British launcher was scrapped. So by 1974 we were into Act IV, and the British programme was reduced to a succession of small Ariel satellites, to be launched by US Scout rockets. In Act V, the launches ceased to be free, the manufacturing costs of the satellites escalated, and so the Ariel series itself was abandoned with Ariel 6. As Prospero aptly remarked, 'Our revels now are ended'.

But not quite: there remains the hope of small home-made, do-it-yourself satellites, which can 'thumb a lift' on launch vehicles with power to spare, or be pushed out from the Shuttle orbiter. The University of Surrey showed the way with Uosat (1981-100B), designed for use by schools.

Salyut and Soyuz

Though Britain rarely supplies observers with new satellites, plenty are still being sent up from the USSR — 75% of the launches and 90% of the tonnage during the 1970s. The Russian manned space flights have relied on the Soyuz spacecraft, which carries a crew of two and has been used as a ferry to the Salyut 'space station'. In its current version Salyut is about 22 m long with a maximum diameter of 4 m and a mass of about 19 tons: three large panels covered with solar cells provide power. Fig 22 shows Salyut and Soyuz joined. Salyut 1 (1971-32A) was in orbit for six months, and later Salyuts have operated for longer periods and have received numerous visitors.

Fig 22 *Salyut 'space station', with Soyuz spacecraft attached at left and a Progress unmanned craft at right. (At Soviet 'Exhibition of Economic Achievements', 1981.)*

They have arrived two by two in Soyuz spacecraft, which can dock at either end of the Salyut; and they have sometimes returned in an earlier Soyuz, which was already docked. Salyut 6, launched in September 1977, remained in orbit for more than five years and was occupied by two, or occasionally four, astronauts for most of its time in orbit. When its operational life ends, a Salyut is commanded to descend over the Pacific Ocean, where it burns up in the atmosphere.

The first Soyuz flew as long ago as 1967, and there were 46 Soyuz launchings up to July 1982. The Soyuz is 10 m long, with a maximum diameter of 3 m and a mass of about 6 tons. It is made in three sections: a large 'living room'; a re-entry capsule (in the middle); and an instrument section. The first and third of these are jettisoned before re-entry.

Both Soyuz and Salyut are bright objects in the night sky. They are usually to be found circulating at a height between 200 and 300 km in orbits inclined at 51° to the equator. As they pass overhead, Soyuz is usually of magnitude 1, while Salyut is of magnitude 0 and sometimes shines more brightly if its solar panels happen to reflect light to the observer. Usually a Soyuz docks with Salyut within about 24 hours of launch, and so it is not easy to observe a free Soyuz. But if you regularly observe the current Salyut, you occasionally see a new Soyuz nearby, about to dock with the Salyut. After docking, the combined Salyut-Soyuz (or occasionally the Soyuz-Salyut-Soyuz, if there are two docked) is even brighter than the Salyut alone, usually of magnitude -1.

Weather satellites

Taking photographs of the Earth from space to show the cloud cover (or the lack of it) was one of the obvious practical uses for satellites, and the first weather satellite, the American Tiros 1, was launched in 1960. The Tiros series has been followed by many other weather satellites registering the visible scene and also that recorded at infra-red and other wavelengths.

The most easily observed of the weather satellites are the Russian Meteor spacecraft, of which 39 had been launched up to August 1982. The Meteor is about 5 m long and 1.5 m in diameter with two large solar panels, and usually enters a nearly circular orbit inclined at 81° to the equator at a height near 900 km, after separation from its rocket, which is 3.8 m long and 2.6 m in diameter. The Meteors and their rockets are expected to have lifetimes of about five hundred years, so they provide a 'reserve' of observable objects if other launches should cease in the future. The Meteors and their rockets appear at about magnitude 5 on an overhead transit. Their orbits, though not usually of great interest, occasionally pass through 'resonances' when the track over the Earth repeats, usually after 14 revolutions for the Meteors, and such resonant satellites are often selected for intensive observation.

The US weather satellites, under the names Tiros, Nimbus, NOAA, and DMSP, have in recent years usually entered Sun-synchronous orbits at an inclination of 98°, chosen to ensure that the orbital plane remains fixed relative to the Sun, so that the satellite takes photographs at approximately the same local time on each revolution. For example, the satellite may have midday and midnight views on south-and-north-going transits, or sunrise and sunset views. (Some of the Meteors are also in 98° orbits.) Visually the US weather satellites are generally not so bright as the Meteors, but they are specially observed whenever their orbits pass through an 'interesting condition' for geophysical researches.

There are also some weather satellites, such as the European Meteosat, in stationary orbits at heights near 36000 km: these are not visible, except to observers who have access to a very large telescope.

'Big Bird'

The brightest US satellite to be seen at high latitudes during the 1970s and early 1980s was the *low-a*ltitude *s*urveillance *p*latform, or LASP, which has the rather incongruous pet name of 'Big Bird', in recognition of its size, a cylinder 15 m long and 3 m in diameter, with a mass of 12 tons. A later version is known as KH 11. These satellites pursue their task of photographic and other reconnaissance from low orbits at heights between about 160 km and about 500 km. A satellite left alone in such an orbit decays after a few days or weeks; but the LASP has an on-board motor, to adjust the orbit every few days and restore the status quo. The orbital inclination is 96°, chosen so as to ensure that the local time of the photographs remains nearly the same day after day.

In recent years the LASPs have been kept in orbit for more than 12 months. Being so low, they are very bright, magnitude 0 or even -1 on an overhead transit, and move quickly across the sky. But they are not always easy to observe because they are apt to enter eclipse and because the orbital changes make accurate prediction difficult, though prediction becomes easy if the manoeuvres are designed to keep the orbital period constant.

Comsats

The favourite location for communications satellites is in an eastbound equatorial orbit at a height of 36000 km, where a satellite moves at the same angular rate as the Earth. In this 'synchronous' orbit, the satellite appears to remain stationary relative to the Earth. Cosily congregated in their stationary ring, these are socially the most important of satellites. By bringing distant sporting and cultural events to the television screen, they foster the 'Olympic spirit' of peace and goodwill between nations. However, they also bring religious and ideological hatreds instantly to the screen, give publicity to evil people and drive good-natured viewers to cynicism and despair. Direct-broadcast television may well cause further trouble by subjecting one country to subversive propaganda from another.

Though the synchronous comsats can throw disturbing images on the television screen, they are poor specimens for observing, being of magnitude about 14 and invisible to the eye without the aid of a large telescope. Also, being stationary, they are difficult to distinguish from the nearby stationary stars, and are best recorded by photographs with a long time-exposure, when the stars trail and the satellite is distinguishable as a point.

Nobody would expect to be able to observe synchronous equatorial comsats, but you might think you would do better with the Russian Molniya communications satellites, of which over a hundred have been launched. Their orbits are at inclinations near 63° and initially have perigee heights near 400 km and apogee heights near 36000 km. But the perigee always stays well in the southern hemisphere, so that the satellite is always at a great height when north of the equator: it is designed as a northern-hemisphere comsat. Occasionally a Molniya or its rocket can be 'caught' by a northern-hemisphere observer during the two or three weeks before decay when the apogee height has decreased below 10000 km and observation is possible. But this is difficult because the decay is very rapid; so Molniyas generally show themselves only to southern-hemisphere observers, though flashes from their flat solar panels can sometimes be seen, even at a distance of 40000 km.

With apologies

Many of the 5000 satellites in orbit have not been mentioned at all in my rapid run-around. I am sorry not to have mentioned the Ablestar rockets, smaller brothers of the Agena, 4 m long and 1.2 m in diameter, several of which remain in orbit and are reasonably bright. My apologies also to the multitude of faint and fragmentary objects which may be doing good work for those who launched them, but cannot be called prime specimens for observation. They need not feel too neglected, because observers are great individualists, and some like to look at as many satellites as possible: so even the outcasts, like fragment 1972-58AD, are being observed sometimes, by someone, somewhere.

The Space Shuttle

This chapter can appropriately end on a forward-looking note, with the Space Shuttle, or Space Transportation System (STS), which itself opens a new chapter in space launching. The idea of providing a regular shuttle service between Earth and space — rather like the Boston-New York-Washington air shuttle, though not so frequent — arose because of the great expense of throwing away huge rockets with every launch. A reusable cargo-carrying spacecraft, taking off like a rocket and landing like an aircraft, could greatly reduce the cost of putting satellites into orbit. The design and development work went on through the 1970s, leading to the layout already illustrated in Fig 6, and the first Shuttle orbiter, Columbia (STS 1), was successfully launched from Cape Canaveral on 12 April 1981. The solid-fuel boosters were recovered from the ocean, according to plan, and two days later the Orbiter itself made a perfect landing in California. The stage was set for many more such flights. Columbia flew again in November 1981, and its third flight was in April 1982.

The STS Orbiter, Fig 23, is 37 m long with a wing span of 24 m. Its operational weight is 68 tons and it can carry payloads ranging between 14 and 29 tons, depending on the orbital inclination and altitude. For the first few years the launches will be at inclinations between 28° and 57°, and the Orbiter will usually enter a nearly circular orbit at a height between 200 and 500 km. Seen at a distance of 500 km it is of magnitude 0, and at 300 km magnitude -1, or sometimes even brighter when the full wing area reflects sunlight.

The European Space Agency has designed and built the 'Spacelab' to fit into the cargo bay of the Shuttle Orbiter, and most of the scientific experiments will be conducted using this facility. The first Spacelab flight, with European scientist-astronauts, is scheduled for 1983. The Orbiter will also eject free-flying scientific satellites, and although the Orbiter itself will only stay in space for a week or two at a time, it should (whenever necessary) be able to recover some of these satellites from orbit, months or years after their original launch. Such ejected satellites should be visible as fainter objects separating from the Orbiter, or possibly flying in formation with it. Observers are looking forward to this rather fascinating spectacle.

Fig 23 *The Space Shuttle Orbiter, which is 37 m long with a wing span of 24 m, and can carry payloads of up to 29 tons.*

4
To See
or Not to See

Bright star, would I were stedfast as thou art.

John Keats, *Sonnet* (1819)

A bright satellite moving across the sky looks as steadfast as any star — until it suddenly disappears on entering the Earth's shadow. In the winter most of the satellites passing over during the night are in eclipse and quite invisible to us. How do you know whether a satellite can be seen or not? Why is it that a particular satellite sometimes cannot be seen for several months? This short but strong chapter is intended to give at least a partial answer to such queries, by explaining how orbits swing around in space, and flip in and out of the Earth's shadow to produce cycles of visibility and invisibility.

Sunshine and shadow

A satellite can only be seen if it is illuminated by the Sun against a dark sky background: the satellite must be in sunshine and the observer well in shadow. To make this statement precise, consider the height of the Earth's shadow vertically above the observer. If the sky is to be dark, the Sun should be at least 10° below the horizon: the shadow height is then at least 100 km. And if the satellite is to be illuminated, it must be higher than the shadow. Faint satellites can therefore only be observed when the shadow height is greater than 100 km but less than the satellite's height, which may be as low as 200 km. Consequently, evening observations of low satellites are usually possible only during the hour or two while the shadow is climbing from 100 km to the satellite's height. (Here and later I shall refer only to evening observations, because most people are awake in the evening and asleep in

45

the early morning: but the situation in the morning is similar, except that the time sequence is reversed and observations have to be made before the onset of morning twilight.)

Because of the Earth's shadow, a low satellite can on any particular day normally be observed only from two narrow bands of latitude, one in each hemisphere. This is one of the gravest limitations of optical tracking, because orbits can be determined more accurately if observations are uniformly distributed round the Earth. Fig 24 illustrates the visibility conditions during the northern summer for a near-polar satellite which has the Sun nearly in its orbital plane. The satellite can only be observed easily over about 20% of its orbit.

The Earth's shadow is shown as a sharp-edged cylinder in Fig 24: this is a useful simplification, but the shadow is really a fuzzy and variable slender cone more than a million kilometres long. The edge is fuzzy in the penumbra, where only a part of the Sun's disc is visible; and variations occur because of the variable height of the clouds at the (varying) point where the Sun's rays are grazing the Earth. Sometimes

Fig 24 *If a satellite is to be visible, it must be illuminated by the Sun against a dark sky. This diagram shows how observations can usually be made on two sections of the orbit, one in the northern hemisphere and one in the southern. At P the shadow height PS is about 6000 km.*

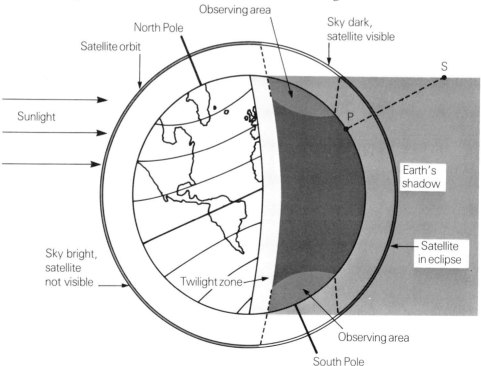

a satellite will enter eclipse quite suddenly; but often the process of eclipse will take 10 or 20 seconds, or even longer if the satellite's path runs nearly parallel to the edge of the shadow.

The height of the Earth's shadow is absolutely vital for satellite observing. For example, at the point P in Fig 24, the shadow height PS is approximately equal to the Earth's radius or about 6000 km. So the only satellites visible overhead at P would be more than 6000 km high and therefore probably very faint and difficult to observe.

The height of the Earth's shadow above you at any particular time in the evening varies greatly according to the time of year. Fig 25 is a chart showing the height of the shadow throughout the evening at latitude 50°N for any date. The time scale is the local time. For observers near London this would be Greenwich Mean Time (GMT); but for every degree of longitude west of London, 4 minutes must be subtracted, so that in south-west Ireland (10°W) local time is GMT minus 40 minutes; and at longitude 75°W (roughly the longitude of Ottawa or New York) local time is GMT minus 5 hours, ie United States Eastern Standard Time.

The best way of understanding Fig 25 is to use it. Suppose the date is March 2 and you are near latitude 50°N. Proceeding along the horizontal line to the right of 'Mar.2', you find that sunset is at about 17.45 local time (that is, 17 hours 45 minutes), and twilight lasts nearly an hour. So the sky won't be dark enough to observe faint satellites until about 18.40 when the shadow height is 100 km. By 19.10 the shadow height has reached 200 km, and after that you won't be able to see any satellite lower than 200 km overhead, though you can still observe such satellites in the west, where the shadow height is lower. By 19.30 the shadow height has risen to 350 km, and no satellite lower than this can be seen overhead. By 21.00, or 9 pm, the shadow height has reached more than 1000 km: many of the brighter satellites will be in shadow as they pass over, but there are still plenty at heights between 1000 and 2000 km to be seen. And so on.

To use Fig 25 in another way, suppose that a satellite is in a circular orbit at a height of 400 km. The '100' and '400' curves in Fig 25 show that on January 1 it can be seen overhead at 50°N only between 17.12 and 18.18: the available time interval remains at about one hour from January until mid-April, when it begins to lengthen. Between 20 May and 20 July the satellite would be visible at any time during the night: observing in summer, with no eclipse to worry about, is easier and more pleasant, except at high latitudes, where the sky is never dark. In June any satellite higher than 300 km is always above the Earth's shadow at latitudes greater than 50°N.

The observing time available is really greater than Fig 25 might suggest, because a satellite need not be exactly overhead: it can be observed in the dark eastern sky a little before the end of twilight and also, for a satellite at 400 km height, at longitudes up to 10° to the west of the observer, ie at times up to 40 minutes later than Fig 25 indicates. Making an addition of half an hour for a satellite at 200 km height

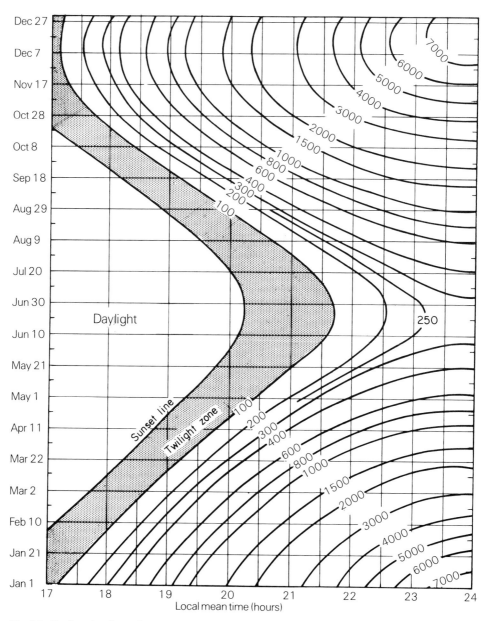

Fig 25 *Shadow height at latitude 50°N. The numbers on the curves give the height in km of the Earth's shadow at latitude 50°N at any time in the evening and any date during the year. Suppose, for example, that a satellite passes over an observing station near latitude 50°N at 21 h 30 m local time on 29 August. The diagram gives the shadow height as 500 km, so the satellite will be in shadow unless its height exceeds 500 km.*

48

and about an hour for 400 km, we can say that at latitude 50°N over the greater part of the year (excluding May, June and July) the possible observing time for a faint satellite is about 1 hour if its height is 200 km, and about 2 hours if its height is 400 km. When only one hour is available, a low satellite may be difficult to observe at all on some evenings, because it may come past a few minutes before the sky is dark and then for a second time $1\frac{1}{2}$ hours later, when it is too low in the west.

Nearer the equator, the shadow height increases more quickly than shown in Fig 25, and the time available for observation is shorter. At the poles, however, the shadow height never exceeds 600 km, and polar satellites above that height can not only be seen every day during the polar winter but can also (weather permitting) be seen on every revolution. If you want to be near the Sun, you should go to the north or south pole.

This is a good moment for me to make my apologies to readers in the southern hemisphere. Regrettably, but unavoidably, my diagrams and explanations have a northern bias.

Prediction

Fig 25 tells you whether a satellite is in sunshine or shadow when it passes overhead at a latitude near 50°N; but of course you also need to know the time of its overhead passage. The plane of the orbit remains nearly fixed in direction in space as the Earth spins beneath, and you can only observe the satellite overhead when the Earth's rotation carries you across the plane of the orbit; this may happen at any time of day, and not necessarily while the shadow height is favourable. Forecasting when a satellite is going to pass by is called *prediction*, and fortunately satellite predictions are much more reliable than the predictions of weather, future events, etc., that have given the word 'prediction' rather a bad name. For satellites of long life, little affected by air drag, prediction is easy and accurate, once the satellite has been spotted on two successive nights.

For simplicity, consider first a polar orbit, so that the inclination to the equator is 90°. Suppose that you see it cross your latitude going north 8 minutes later on Tuesday than on Monday. Then, if its orbit is little affected by air drag, you can safely say it will be 8 minutes later on Wednesday than on Tuesday, 8 minutes later still on Thursday, and so on, right through the week.

But the longitude at which it crosses your latitude will also steadily change, for two reasons. First, even if the satellite arrived at the same time each evening, its longitude would be 1° further west (or, strictly, 0.986°). This is because, relative to a fixed direction in space, the Earth rotates through 361° (or, strictly, 360.986°) in 24 hours. All the stars, and polar satellites (which stay in a plane fixed in direction in space), therefore appear to be 1° further west than at the same time on the previous night. The Earth spins through 360° in 23 hours 56 minutes, and spends the remaining 4 minutes 'catching up' on the small angle needed to bring it to the same direction relative to the Sun: since the Earth completes its journey round the Sun in 365.25 days, this angle is 360°/365.25 = 0.986°.

The second reason for a change in longitude is the different time of arrival each day. If the satellite reaches your latitude 8 minutes later each evening, the Earth has spun through nearly 2° in the 8 minutes, so the satellite will appear 2° further west for that reason.

Thus a polar satellite which appears 8 minutes later each evening will be 3° further west in longitude each day, 1° arising from the first effect and 2° from the second. So it might cross your latitude 4° east of your longitude on Monday, 1° east on Tuesday, 2° west on Wednesday, and so on. After about a week it will be too far to the west to be seen, though the previous transit may then become visible in the east.

So far we have only looked at polar orbits, and the plane of the orbit then stays fixed in direction in space. But if the inclination *i* differs from 90°, the gravitational pull of the Earth's equatorial bulge makes the orbital plane rotate slowly about the Earth's axis. If the satellite travels towards the east, as most satellites do, the orbital plane rotates towards the west, as shown in Fig 26. Suppose it rotates at 4° per day, a fairly typical value. Then a satellite which crosses latitude 50°N at the same time on two successive days will be 1° further west on the second day because of the Earth's rotation, and a further 4° west because of the rotation of the orbital plane. Altogether the satellite will be 5° further west.

Fig 26 *The orbital plane of a satellite does not remain quite fixed in direction in space, but rotates slowly about the Earth's axis, to the west if the satellite is travelling eastwards. The rate of rotation can be as large as 8° per day (see Fig 27).*

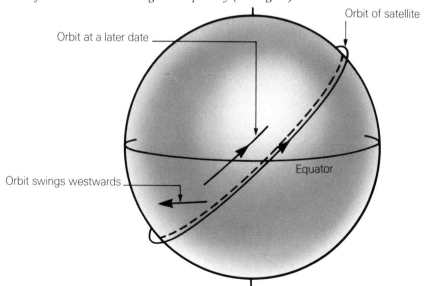

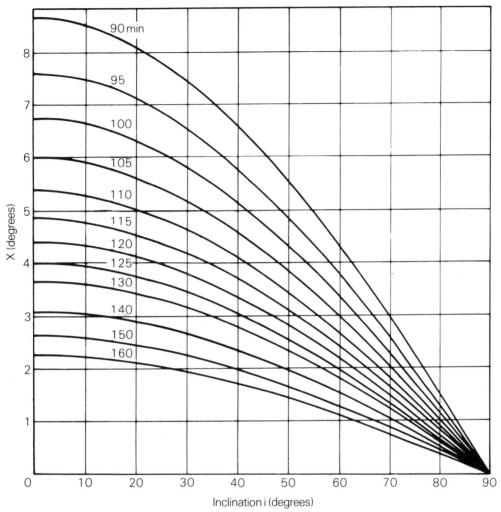

Fig 27 *The daily shift to the west, X, of a satellite's orbital plane. The numbers on the curves give the orbital period in minutes. The curves are for near-circular orbits. When the inclination i exceeds 90°, ie the satellite travels westwards, the orbital plane swings to the east, and X has the numerical value appropriate to an inclination of 180° − i.*

The rotation of the orbital plane is vitally important in prediction, and Fig 27 shows how far the plane rotates to the west each day, for orbits of any inclination and with periods of 90, 95, 100 . . . minutes. If the orbital period is 100 minutes, which is near the average among satellites usually observed, the daily shift to the west, X, is near 6° for an inclination of 30°, rather over 3° when $i = 60°$, and of course zero when $i = 90°$.

Adding in this third effect, we see that a satellite which crosses your latitude N minutes later on successive nights will be at a longitude $(1 + X + \frac{1}{4}N)$ degrees further west.

As an example, take a low satellite with an inclination of $73°$ and an orbital period of 89.5 minutes, similar to many of the bright Cosmos reconnaissance satellites. For a satellite with $i = 73°$ and period 89.5 minutes, Fig 27 gives $X = 2.5°$. Since each revolution is $\frac{1}{2}$ minute short of 90 minutes, 16 revolutions will be 8 minutes short of 24 hours, so the quantity N is -8. The satellite therefore appears 8 minutes earlier on Tuesday than on Monday and crosses your latitude at a longitude $1.5°$ further west (on substituting $X = 2.5$ and $N = -8$ in the formula at the end of the last paragraph). On Wednesday it will be another 8 minutes earlier and another $1.5°$ further west, if we ignore the effects of air drag.

As this example shows, doing your own predictions is quite easy once you have seen the same satellite on two successive nights. The real problem is to know when

Fig 28 *Successive transits of a satellite with a 90-minute orbital period are at intervals of about $22\frac{1}{2}°$ in longitude. The example shown is for a satellite with an inclination of $65°$.*

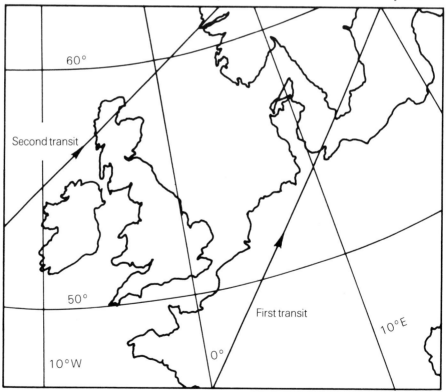

to look for the satellite to start with, and then to decide whether you have seen the same satellite on the next night. So do-it-yourself prediction is a hit-or-miss procedure, and in practice it is much more efficient for a computing centre to keep details of the orbits of chosen satellites and issue predictions to observers.

If the satellites were not affected by air drag, the prediction centre would have an easy job and could send out predictions for months ahead. But every satellite affected by air drag suffers a daily decrease in orbital period, and the decrease varies each day, depending on the irregular variation of air density, which itself depends on the unpredictable vagaries of solar activity, as we shall see in Chapter 10. In practice, therefore, predictions are usually issued weekly: the predicted times at the end of the week will be accurate to a few seconds for high satellites, but may be in error by up to 10 minutes for a satellite with perigee height less than 200 km. The procedure for prediction is just the same as for drag-free satellites, but the predictors have to make a choice of the rate of decrease of orbital period, and use a shorter period each day, so that the quantity N decreases each day.

Predicting from one day to the next is obviously necessary, but often a satellite can be seen on two successive revolutions, so prediction one revolution ahead is also needed. If you know the orbital period T (minutes) of the satellite, the time of the second transit of your latitude is merely T minutes later than the first; and since the Earth turns through $1°$ of longitude in about 4 minutes, the longitude of the second transit is about $\frac{1}{4}T°$ further west than the first. Thus a satellite of period 90 minutes has its second transit about $22\frac{1}{2}°$ west of the first, as shown in Fig 28. If you want to be more exact, you should allow for the rotation of the orbital plane, etc. After 16 revolutions, the satellite would be $(361 + X)$ degrees further west, where $X = 3.7$ if $i = 65°$, from Fig 27: so after one revolution the longitude would be 364.7/16 degrees, or $22.8°$, further west.

Spells of visibility

We have already seen how the longitude of a satellite tends to drift each night, usually to the west, and this means that most satellites have rather short spells of visibility, generally from one to three weeks, with a repeat performance a few months later.

To illustrate what happens, consider a satellite in a nearly circular, nearly polar orbit at a height of about 500 km. Suppose that initially it is passing over going north at midnight. The shadow height will then be greater than 500 km (unless the season is near midsummer), so the satellite is in shadow. Gradually the orbital plane will swing westwards, at $(1 + X)$ degrees each day relative to the Sun, and will eventually reach the point where the shadow height is less than 500 km. The satellite will then enjoy a spell of evening visibility for a few weeks. As the orbit swings further to the west, the satellite is lost in the glare of the setting sun. After that it will be passing over in the daytime for many weeks. Then it moves into the morning twilight, and, after a spell of morning visibility, once again enters eclipse. The cycle is complete, and starts again.

There is an equal chance of a southbound rather than a northbound transit, and the southbound transits follow the same sequence, separately from the northbound transits.

If the orbit of the satellite is strongly retrograde, with an inclination greater than 100°, the orbit swings to the east rather than to the west, because $(1 + X)$ is negative, and the normal cycle of visibility is reversed, with morning visibility followed by daytime transits, then evening visibility and midnight eclipse.

Between the normal and reverse cycles there is the possibility of a Sun-synchronous orbit with $X = -1°$, so that the orbital plane remains fixed relative to the Sun. An orbit of 90 minutes period is Sun-synchronous when its inclination is 96°, as with the low-altitude LASP satellites; while a 100-minute orbit is Sun-synchronous if its inclination is 99°, as with the DMSP weather satellites.

This gives the general picture; but just how long does a spell of evening visibility last? This question is so important that I shall attempt an oversimple reply, which should usually give the right impression and only occasionally mislead. We have seen how a satellite that crosses your latitude N minutes later each day comes over at a longitude $(1 + X + \frac{1}{4}N)$ degrees further west each day. So it will arrive at the *same* longitude each day if $N = -4(1 + X)$, that is, if it comes over $4(1 + X)$ minutes earlier each day, where X is given by Fig 27. The length of a spell of evening visibility is the time taken for the orbit to move through the evening visibility time, as indicated by Fig 25 with half an hour or an hour added, as explained earlier. Since it comes over at the same longitude if $4(1 + X)$ minutes earlier each day, the spell of evening visibility lasts for $V/4(1 + X)$ days, where V is the length of the evening visibility time in minutes.

Again an example may help. For a satellite at height 400 km, as we saw earlier, the visibility time is about two hours for much of the year, so $V = 120$ minutes and the spell of visibility lasts for $30/(1 + X)$ days. If the orbit is polar, X is zero and the satellite is visible for about 30 days. If the inclination is 70° and the period about 92 minutes, $X = 2.8$ from Fig 27, so the satellite is visible for $30/3.8 = 8$ days. For a satellite at 600 km height, V averages nearly 3 hours and the spells of visibility last nearly 50% longer, that is, 45 days or 12 days instead of 30 days or 8 days.

The formula $V/4(1 + X)$ becomes inaccurate when the spell of visibility exceeds about a month, because the shadow height itself may change appreciably in that time. This inaccuracy is worst when X is near -1.

The formula gives an idea of the spell of visibility if the satellite comes over *either* on northbound *or* on southbound crossings, as happens with near-polar orbits. If the orbital inclination is less than or only a little greater than the observer's latitude, however, the northbound and southbound sets of transits tend to merge and the visibility period can be more than twice as long as might be expected.

After enjoying a spell of visibility, how long do we have to wait before seeing the satellite again? A satellite comes at the same longitude each day if it is $4(1 + X)$ minutes earlier each day. It will be visible at the same time of day again when it is 24 hours (1440 minutes) earlier, that is after $1440/4(1 + X)$ days or $12/(1 + X)$ months. So if $X = 3$, the interval is three months. Fig 29 shows the average time interval between successive spells of visibility on northbound (or southbound) transits, and also gives the fraction u of the time interval which elapses between northbound and southbound spells of visibility.

Fig 29 shows for example that a satellite with inclination 70° and period 100 minutes would be visible on northbound transits at average intervals of about $3\frac{1}{2}$ months; and since $u = 0.36$, a southbound spell of visibility occurs about $1\frac{1}{2}$ months after the northbound spell. These figures are averages over the year: the intervals tend to be slightly smaller in the spring and greater in the autumn. An indication of seasonal variations from the average is shown by the broken lines in Fig 29.

Fig 29 shows that the interval between successive spells of visibility can be as little as five weeks for near-equatorial orbits. This does not mean that equatorial latitudes are particularly favoured, for the spells of visibility themselves are correspondingly shortened.

For exactly polar orbits, or for very distant orbits, the interval between successive northbound spells of visibility is one year, with southbound transits halfway between.

If a satellite is in a strongly retrograde orbit with $(1 + X)$ negative, the formula $12/(1 + X)$ still holds, but the answer is negative: this expresses the fact that the cycle is reversed.

These 'rules of thumb' for visibility periods provide a fair guide for nearly circular orbits, but exceptional situations can arise. For example, between 20 July and 28 October, the shadow height at latitude 50°N reaches 200 km about 2.4 minutes earlier each day, with very little variation. So a satellite 400 km high which is suitably placed for observation on 20 July would continue to be visible each night until after 28 October, if its orbit is such that it would come over at a constant longitude when it is 2.4 minutes earlier each day. The formula already given shows that this happens if $4(1 + X) = 2.4$, which gives $X = -0.4$. Fig 27 shows that $X = 0.4$ (for period 92 minutes) when $i = 87°$: therefore $X = -0.4$ when $i = 93°$. This orbit can safely be called exceptional, since no satellites of inclination 93° have yet been launched.

Highly eccentric orbits provide another exception to the rules. By bad luck some such satellites may manage to avoid being seen even when they are theoretically in a spell of visibility. If the orbital period is, say, four hours and perigee is near your latitude, it can happen that the satellite is on the other side of the world during the two hours or so of the evening when you have a chance to see it, and this bad luck may continue for a week or more. On the other hand you may find that you can see bright satellites in highly eccentric orbits while they are near apogee over a distant

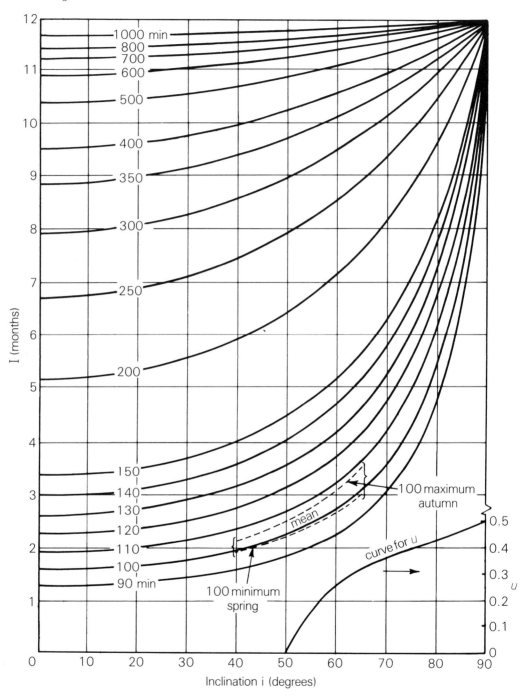

Fig 29 *The average time interval I between successive northbound spells of visibility. Numbers on the curves give the orbital period in minutes. The values of I are average over all seasons: in spring I may be slightly smaller and in autumn greater. An example of the extreme seasonal limits is shown by the broken lines.*

The time between northbound and southbound spells of visibility at latitude 50°N is ul, where u is shown in the subsidiary graph.

country, when they are not 'officially' having a spell of visibility at all in your own area.

You can observe satellites without knowing anything about predictions and spells of visibility. But the variations in satellite visibility seem capricious if you remain ignorant, and most observers like to have an idea of why and how the satellites flip in and out of shadow. I hope this chapter has given at least a rough guide.

5
With your own Eyes

And people without the reflector, with their opera glasses,
will be able to see sufficiently well.

E.E. Hale, *The Brick Moon* (1871)

Would you like to observe satellites visually, using binoculars, opera glasses, or just sharp naked eyes? If so, this is the chapter for you.

Physically, satellite observing is quite an easy pastime to pursue. There is no need to tramp for miles through mud, or crouch for hours in an uncomfortable hide, as some naturalists do. Instead you go out in the garden about two minutes before the satellite is due, recline in a deck-chair, look for the satellite, note the time when it passes particular stars, and come indoors again after enjoying a few minutes of fresh evening air. The air can be all too fresh if propelled by a freezing wind, but you are not exposed for long, and rain or cloud entitles you to a night off. Usually you only observe during the two or three hours after twilight ends (though you can always observe before dawn too if you are really keen). In England about two nights a week are clear, on average, and you may try to observe four or five satellites on each.

Observing is not an expensive pastime: a golfer, for example, needs far more expensive equipment than an observer, and has to pay subscriptions or green fees, whereas the satellites are there for anyone to observe, free of charge. The basic minimum of equipment is an accurate wrist watch and a keen eye. But if you are serious, you need a good stopwatch (a quartz digital wrist watch with stopwatch facility is adequate), a cheap pair of 7 × 50 binoculars, which you may already have, and Norton's *Star Atlas* (£6 in 1982). As with any pastime, you can spend more if you wish. Next on the shopping list would be the large Bečvář star atlases, the *Atlas Borealis* and *Atlas Eclipticalis* (£20 each in 1982). You also need to be able to identify a few bright stars like the Pole Star, Vega, Capella and Arcturus: for the rest you can use the star atlas. (If you are in the southern hemisphere, the *Atlas Borealis* should be replaced by the *Atlas Australis*, and the Pole Star, etc should be replaced by the Southern Cross, Sirius, etc.)

In practice, if you become a serious observer, you will probably be supplied with the Bečvář star atlases; Norton's atlas, binoculars and stopwatch are the essentials.

Although observing is physically and financially undemanding, you will need all your powers of alertness and concentration to make three accurate observations of a satellite on one transit, as most observers try to do. To make three accurate observations of two unconnected satellites in immediate succession, or, worse still, three such satellites, needs powers of memory, organization and efficiency that I would call superhuman if I were not sure that someone would claim to have done it.

In this chapter I shall start with the simplest form of observing, with naked eye and wrist watch, and proceed to the more accurate and more satisfying methods made possible by using binoculars. It is worth saying at the outset that binoculars with a field of view smaller than 4°, and astronomical telescopes, are not recommended for beginners.

With naked eyes

Predictions of bright satellites can usually be found in some daily newspapers: in Britain the *Guardian* has for many years given predictions next to the weather map, usually for four or five satellites. To make your observation, you need to go out, wearing a digital wrist watch (and clothes if you wish, but that's not essential), about three minutes before the predicted time, which may be in error by a minute or two. You should go to the darkest place you can find where you can see the region of sky the satellite will cross. First, you should look for some stars: if you can't see any, the sky has suddenly clouded over (unless it was cloudy all the time and you didn't realize) or the place you have chosen isn't dark enough; so you give up. But if you can see the stars, you are in business, and you should at once start looking for a moving star — the satellite — and go on looking for about five minutes. Of course you may not see the satellite. Perhaps it was too faint, perhaps the prediction was in error, or perhaps you looked in the wrong direction. Whatever the reason, you will have to try again. When, at your first attempt or your umpteenth, you do see the satellite, follow it until it crosses the (imaginary) line between two bright stars, A and B say, reasonably close to each other. As it crosses the line, estimate its fractional distance between A and B in tenths: for example, 6 tenths up from A towards B (if B is the higher), as in Fig 30. Then quickly press the light switch on your watch and read the time to the nearest second if it has seconds digits (or the nearest

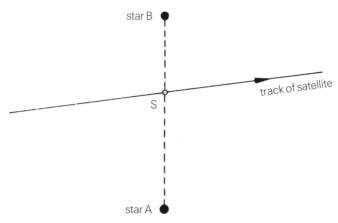

Fig 30 *The basic observation. The satellite S passes between two stars A and B, and you estimate the distance AS as $\frac{6}{10}$ of AB, at the same time looking at your watch, or, better, starting a stopwatch.*

minute if it hasn't). You have now made an observation, though probably not a very accurate one.

Next you must identify the stars you used. You may know them already, for example the two right-hand stars in the 'W' of Cassiopeia, or the two stars forming the right-hand side of the square of Pegasus. But if you can't identify them, you must look them up on a star map if you want your sighting to qualify as an observation. Suppose your stars A and B are α (alpha) and β (beta) Pegasi, the two stars forming the right-hand side of the square of Pegasus. They are 13° apart and the error in estimating distance is usually about $\frac{1}{20}$ of the distance between the stars; so your observation, recorded as '6 tenths up from α towards β Pegasi' could be accurate to 0.6°.

Next there is the timing. If you failed to write down the time, you will probably have forgotten it by now, but if you still remember, you need to subtract perhaps 2 or 3 seconds for the time you took to press the light switch and read the watch, and you also need to check at the next radio or television time signal how many seconds fast or slow your watch was running. After making these corrections, your timing could be accurate to about two seconds.

Here ends the first lesson in observing. Your error, 0.6° and two seconds, is about 20 times greater than skilled observers achieve. But, if you persevere, you can reach a standard approaching that of the best observers within a year or two. So you are much better off than, say, a novice tennis player who, after years of practice, still has scarcely a hope of approaching the standard of the world's best players.

You can obviously do better at your next attempt to observe. Looking down to read your watch after the observation is very inefficient and inaccurate. Most digital watches now have a stopwatch mode, so you would do better to set the stopwatch to zero when you begin looking for the satellite and start it running as the satellite crosses the line between the stars. Then you can stop the stopwatch against a time signal, either the Speaking Clock (if you can afford the phone call) or the next radio or television time signal. Subtract the stopwatch reading from the time at the time signal, and you should obtain the time of the observation correct to about half a second, or even a fifth of a second, if you take care and the stopwatch reads to hundredths of a second, as most of them do. The limitation on your timing accuracy will be your perception of the time when the satellite crosses the line between the stars, and the error from this source would probably be about half a second at best.

You can also obviously do better by using stars less than 13° apart. The stars in the 'W' of Cassiopeia, for example, are about 5° apart, so you could reduce your directional error to 0.25° if you were able to use these, or other pairs less than 5° apart, like Castor and Pollux or the pointers of the Plough — though you should be warned that satellites seem to have an unfortunate habit of not going near any familiar reference stars. If you can use stars 5° apart and stopwatch timing, you should achieve an accuracy of 0.25° in direction and 0.5 second in time.

Your next step forward as a naked-eye observer is to buy a copy of Norton's Star Atlas, which is still the best atlas for satellite observers. Just as an atlas of the Earth has a grid of lines of longitude and latitude, the map of the celestial sphere is ruled by lines of *right ascension* and *declination*. These terms are as familiar to astronomers as steering wheel and brakes to motorists. But not all new observers are astronomers, so I must devote a few pages to explaining right ascension and declination, and the celestial system of timing, called *sidereal time*.

The celestial sphere

If you want to define the positions of the stars, you have to go back to the Earth-centred medieval attitude: you must imagine all the stars are little lights pinned on to an enormous celestial sphere centred at the Earth's centre.

The main features of the celestial sphere are illustrated in Fig 31. The points where the Earth's axis cuts the sphere are, not unnaturally, called the celestial poles, and of course the Pole Star lies within 1° of the north celestial pole. Similarly the line where the plane of the Earth's equator cuts the celestial sphere is called the celestial equator. *Declination* is the angle in degrees measured north or south from the celestial equator: it is the heavenly latitude, and quite analogous to latitude on the Earth. Lines of declination at 20°, 40° and 60° North are shown in Fig 31.

To complete the picture we need the heavenly longitude — namely the *right ascension*. Just as longitude on the Earth can be measured east from the Greenwich meridian, so right ascension is measured east from a fixed point in the heavens which is called the first point of Aries and denoted by the symbol ♈. This curious sign represents the horns of a ram, a relic from the animal-dominated system of zodiacal

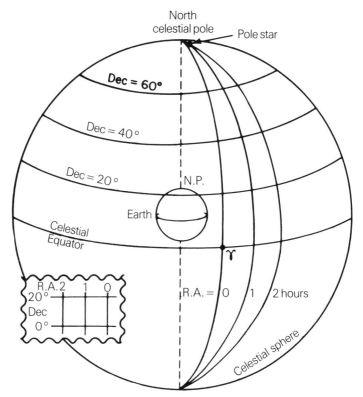

Fig 31 *Celestial sphere, centred on the Earth, as seen from the outside, showing right ascension (RA) and declination (Dec). The inset diagram shows RA and Dec as seen from the Earth.*

signs. The only oddity about right ascension is that it is usually measured in hours and minutes instead of degrees: one hour is equivalent to 15 degrees, and four minutes are equivalent to one degree. Right ascensions of 0, 1 and 2 hours are marked on Fig 31. The hours-and-minutes notation can be rather a nuisance, especially when the hours are indicated by roman numerals, as in Norton's Star Atlas, but it has advantages too, as we shall soon see.

In Fig 31 we are viewing the celestial sphere from the outside. In reality, we see it from the inside, so that the hours of right ascension, looked at from inside, increase on moving from right to left, as shown in the inset diagram. This is slightly peculiar, but no more so than lines of longitude west on a map of the Earth.

We noticed in Chapter 4 how the Earth makes a complete revolution relative to the stars in 4 minutes less than 24 hours. Consequently the stars return to the same position in the sky after 23 h 56 min, and rise, or pass south of the observer, 4 minutes earlier each night (3 min 56 s earlier, to be precise).

To keep pace with this 4-minute-per-day advance, astronomers use *sidereal time*: this is the time recorded by a clock which is synchronized with normal time on 21 September (autumnal equinox) but gains 4 minutes per day, so that it is six hours ahead on 22 December, 12 hours ahead on 22 March, and catches up a full 24 hours by the next September. The sidereal time for an observer on the Greenwich meridian at 0 h GMT on various dates during the year is given in the Table below.

Greenwich sidereal time for 0 h GMT on the 1st and 16th day of each month

	Sid. time			*Sid. time*			*Sid. time*	
Date	*h*	*min*	*Date*	*h*	*min*	*Date*	*h*	*min*
Jan 1	6	41	May 1	14	34	Sep 1	22	39
Jan 16	7	40	May 16	15	34	Sep 16	23	38
Feb 1	8	43	June 1	16	37	Oct 1	0	38
Feb 16	9	43	June 16	17	36	Oct 16	1	37
Mar 1	10	34	July 1	18	35	Nov 1	2	40
Mar 16	11	33	July 16	19	34	Nov 16	3	39
Apr 1	12	36	Aug 1	20	37	Dec 1	4	38
Apr 16	13	35	Aug 16	21	36	Dec 16	5	37

At intermediate dates, calculate from the nearest previous date given, adding 4 minutes per day.

The values are for a year halfway between two leap years, eg 1982 or 1986, and are liable to an error of up to 2 minutes in other years.

To find your *local sidereal time* you take the entry from the Table for the right day of the year and *add it to your local time*. Your local time is close to ordinary clock time if you are near the prime meridian for your time zone. If not, you need to make a small correction, subtracting 4 minutes for every degree of longitude west. In Britain, for example, Greenwich Mean Time (GMT) is local time in London, but 8 minutes should be subtracted from GMT to give local time in Birmingham, and 13 minutes subtracted in Edinburgh. (If 'summer time' is in operation, an hour should be subtracted from ordinary clock time as well as the correction for longitude.)

To clarify the calculation of sidereal time, I shall give two examples. First, suppose you are at longitude 4°W (Glasgow, Swansea or Plymouth) and the time is 18.10 GMT on November 1. Then your sidereal time is 2 h 40 min (from the Table), plus 18 h 10 min, minus 16 minutes (for the 4°W), which gives 20 h 34 min. Second, suppose you are at longitude 80°W (Toronto, Pittsburgh or Miami), and the time is 19.10 EST on November 23. Your local time is 18.50, because you are 5° west of the prime meridian for EST (75°W). The appropriate value from the Table is 4 h 07 min, found by adding 28 minutes (7 days at 4 minutes per day) to the entry for November 16. So your local sidereal time is 18 h 50 min plus 4 h 07 min, that is 22 h 57 min. A small correction is needed in both these calculations because the Table applies for 0 h GMT on the day in question, and you need to add $\frac{3}{4}$ of a day (3 minutes) in the first example, because it is 18.10 GMT. So the right answer is 20 h 37 min, rather than 20 h 34 min. In the second example, the EST time of 19.10 on November 23 corresponds to just after 0 h GMT on November 24, so another 4 minutes needs to be added, giving the local sidereal time as 23 h 01 min.

Your local sidereal time is important in observing, because it tells you the exact right ascension of the stars due south of you (assuming you are in the northern hemisphere). Thus, in the second example, the stars forming the right-hand side of the square of Pegasus will be due south at Miami, Pittsburgh and Toronto, because these stars have right ascensions very close to 23 hours.

So you can now find the right ascension of any star or satellite that is due south — it is equal to the local sidereal time. Can you also find the declination, if you know the elevation above the horizon? The answer is, 'Yes, easily'. If in Fig 31 you draw a line from the Earth's centre to any point S on the 'Dec = 40°' circle, the line makes an angle of 40° with the celestial and earthly equators. So the line cuts the Earth's surface at a (geocentric) latitude of 40°, and at this latitude a star (or satellite) directly overhead will be at a declination of 40°. There is nothing special about 40°: whatever your latitude, you will find that a star or satellite *directly overhead* is at a *declination equal to your latitude*. If the star or satellite is to the south at an angular distance Z from the point overhead, or zenith, its declination is equal to your latitude *minus Z* (and Z is of course equal to 90° *minus E*, where E is the elevation of the satellite when directly south). If the satellite is north at angular distance Z from the zenith, its declination is equal to your latitude *plus Z*. (If the latitude plus Z exceeds 90°, you should add 12 hours to the RA and subtract the Dec from 180°.)

Suppose that the newspaper prediction gives a satellite as visible at an elevation of 66° due south. The zenith distance is 24°, so the satellite's declination is 24° less than your latitude. If you were at 52°N (London, Rotterdam, Warsaw, Irkutsk or Saskatoon), its declination would be 28°N. If you were at latitude 39°N (Lisbon, Samarkand, Colorado Springs or Washington), its declination would be 15°N. If the time of day and year were such that the sidereal time was 23 h, the satellite would coincide with the upper right-hand star in the square of Pegasus (β Pegasi) if you were at latitude 52°N; and with the lower right-hand star in the square (α Pegasi) if you were at latitude 39°N.

So a knowledge of the celestial sphere and sidereal time allows you to plot in Norton's Star Atlas the position of the satellite relative to the stars if the prediction gives the elevation as it passes due south (or north) of you.

With naked but wiser eyes

The digression about the celestial sphere was provoked by Norton's Star Atlas. With this star atlas available, you can now measure the point where you observe the satellite, and give your observation in right ascension and declination, like a professional. For example, your first observation, 6 tenths up from α to β Pegasi, can be found by drawing a line in pencil between the centres of these stars as printed in Norton and marking the point 6 tenths up. The right ascension and declination are not always easy to measure in Norton because the scales are slightly curved, but the error will usually not be more than about 0.2°. The point 6 tenths up from α to β Pegasi is actually near RA 23 h 01.8 min, Dec 22.66°, so you can make your own measurement and see how close you get. Alternatively, suppose your observation is 1 tenth up from the lower to the upper star in the right-hand arm of the 'W' of Cassiopeia, 1 tenth from α to β Cassiopeiae. You can measure it in Norton and see how close your result is to RA 0 h 34.9 min Dec. 56.55°. Although I have quoted these positions quite accurately, you have to remember that your judgment of the 'line between two stars' varies, depending on their elevations and orientation. Strictly the 'line' should be a great circle of the celestial sphere, and if the stars are close enough together, your perception of the 'line' between them will be virtually identical with a great circle. That is yet another reason for preferring stars as close together as possible.

If you have a dark observing point, so that you see stars of magnitude 3 or 4, you should be able to make quite an accurate observation of a bright satellite, by timing it as it passes between a pair of stars less than 5° apart, or by estimating the 'miss distance' from a bright star as a fraction of the distance between a visible pair nearby. For example, the upper right-hand star in the square of Pegasus (β Pegasi) has a pair of fainter stars below and to the right which 'point at' β Pegasi and are 1.2° apart. You may judge that a satellite passes β Pegasi at a distance equal to half the distance between the pair, that is 0.6° from β Pegasi. You also have to remember the direction. The 'passing-between' method is recommended to begin with, but this 'appulse method' may have to be used if the satellite does not pass between any suitable pairs of reference stars.

A skilled naked-eye observer with keen eyesight, using stars not more than 2° apart and a good stopwatch, can often achieve an accuracy of 0.1° in direction and 0.2 second in time. But while you are developing this skill, you will probably become impatient at only being able to observe a few bright satellites. So you will go and buy a pair of binoculars. That brings us to the most popular method of observing.

With binoculars and stopwatch

If you are going to buy binoculars for observing satellites, the best choice is 7 × 50, that is, a magnification of 7 times and a lens diameter of 50 mm. These are readily available, sometimes on 'cheap offer' at less than £30 (in 1982). In the dark the pupil of the eye is about 7 mm in diameter, so in order to take full advantage of your visual powers, the lens diameter in millimetres needs to be about 7 times the magnification. Thus 6 × 40, 7 × 50, 8 × 60, 10 × 70 or 11 × 80 binoculars are all suitable. The

7 × 50 are the easiest to obtain, are much more powerful than 6 × 40, and allow observation of satellites as faint as magnitude 8. The larger binoculars are much more expensive, and some observers find them too heavy, so it is wisest to start with 7 × 50.

Armed with 7 × 50 binoculars, you can on a dark night observe a spherical satellite 1 m in diameter out to a distance of about 900 km, as Fig 16 (see page 29) shows. Nearly all the satellites chosen for prediction have dimensions greater than 1 m, and perigee lower than 1000km, so all of these are within your grasp, though possibly not when at apogee if their orbits are appreciably eccentric.

The most serious obstacle to observing faint satellites is in predicting their tracks: once you have latched on to them, the actual observing is fairly easy. Predictions are usually one of three types: (1) approximate 'newspaper predictions'; (2) geographical tracks and heights; or (3) tracks among the stars, as seen from a particular observing site.

The 'newspaper predictions', for bright satellites, are of the form: 'from London, 18.20-18.24 SW 64S SE*'. This means the satellite is visible from 18.20 to 18.24, rises in the SW, reaches a maximum elevation of 64° in the south, and sets in the SE, with the asterisk indicating that it goes into eclipse before reaching the horizon. Most of the satellites predicted are at heights above 300 km, so the prediction should be fairly good for anywhere within 100 km of London. Obviously this satellite will be lower in the sky if you are north of London, and higher if you are south. If the newspaper also gives a prediction for Manchester, you can interpolate between the two predictions if you live somewhere between. You can make a good number of observations from newspaper predictions, and if you send these observations to the prediction centre, explaining that you are keen to do more observing, you should be supplied with better predictions for more satellites on a regular basis. The British prediction centre, at present at the University of Aston in Birmingham, makes predictions for about forty satellites, more than enough to keep an observer busy. But it is no good expressing the wish to observe without showing any results. Thousands wish to observe, but only a hundred or two actually get down to the hard work of doing it. To send predictions to anyone who asked would be quite uneconomic.

If predictions are sent, they will probably be of the second form, geographical tracks. The time and longitude of the satellite as it crosses the equator are supplied, and also a table showing how time, longitude and height vary with latitude as the satellite moves from the equator round its orbit, and an indication whether it is in eclipse. From these predictions you can plot the track of the satellite over the Earth, preferably on a polar stereographic map, Fig 32. You can then measure the ground distance of the satellite from you when it passes directly south or north of you. The predictions also give the height, and, knowing the height and ground distance, you can find the satellite's elevation above the horizon, using Fig 33, for heights up to 800 km, or its 'big brother' Fig 34 for heights up to 4800 km. You can then find the

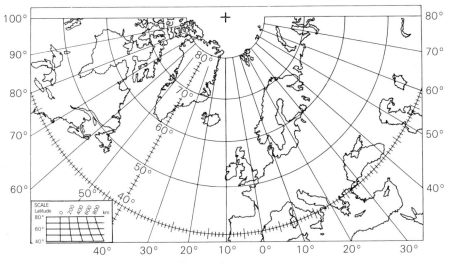

Fig 32 *Map suitable for plotting satellite tracks near Britain.*

Fig 33 *Diagram showing the angle of elevation of a satellite at distances up to 1000 km and heights up to 800 km.*

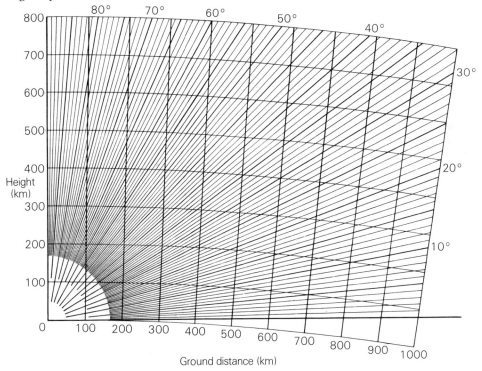

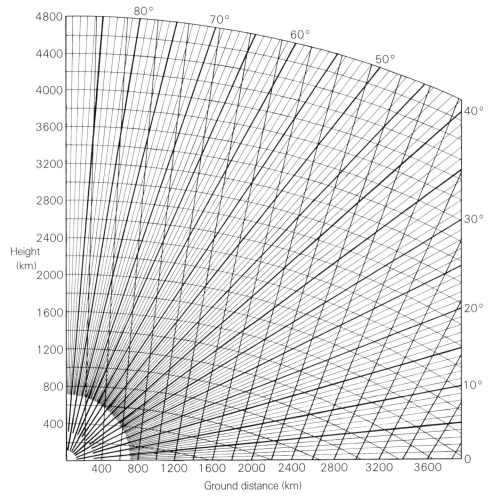

Fig 34 *Diagram showing the angle of elevation of a satellite at ground distances up to about 3000 km and heights up to 4800 km.*

RA and Dec of the satellite as it passes north or south of you, by the method described on page 64. This gives a single position among the stars, and is enough to allow observation.

But if you want to find other points on the track, or if the satellite never passes south or north of you, you need to be able to observe in other directions. In satellite observing the direction is usually given as an *azimuth* angle in degrees: north is azimuth zero; east is azimuth 90°; south is 180°; south-west is azimuth 225°; and so on. From the track diagram you can choose a suitable azimuth, measure the distance to the ground track at that azimuth, and find the elevation from Fig 33 or Fig 34. You then have a predicted azimuth and elevation.

If you wish, you can convert these to right ascension and declination by using a planisphere, or by calculation. The RA and Dec can be found either from the tables given in the British Astronomical Association's Memoir on *Artificial Earth Satellites* (BAA 1961), or by direct calculation from the relevant equations. This is quite easy now, using pocket calculators with trigonometrical functions, and takes about four minutes. If A is the azimuth, E the elevation, and L your latitude, the declination is given by Dec $= \sin^{-1}(\sin L \sin E + \cos L \cos E \cos A)$; while the hour angle h is given by $h = \cos^{-1}[(\cos L \sin E - \sin L \cos E \cos A)/\cos(Dec)]$. The RA is found by subtracting h from the local sidereal time, ie the right ascension of the stars due south. Note that h is positive if the satellite is in the west, and negative if in the east.

On the other hand, to avoid the pain of such conversion, you may prefer to use the azimuth and elevation directly. The rough-and-ready method is to point your deck-chair in approximately the required direction and raise the binoculars to what you think is the right elevation. More accurate methods can easily be devised: if you have a garden, you can mark points on the lawn where the television aerial on the roof is at 30°, 40° or 50° elevation; or you can mount the binoculars on a rotatable wooden frame with an elevation scale; and so on, depending on your temperament and circumstances.

Perhaps you will decide to avoid all the bother by concentrating on naked-eye satellites — detecting them with the naked eye and transferring to binoculars to make the observation. If you become skilled and productive with these or with the geographical-track method, the prediction centre should in due course supply you with 'look angles' giving the track of each satellite among the stars in RA and Dec as seen from your observing site. All you have to do then is to plot the track in Norton's Star Atlas, and go out to observe. Obviously you can achieve far more in observing if the way is thus made easy for you.

If you have obtained predictions relative to the stars by one or another of these methods, the next step is to draw the predicted track, or mark the predicted point, in soft pencil in Norton's Star Atlas. If you have a track rather than a point, you can choose where to observe, and you will probably prefer to observe quite near a familiar star or star-group. Then you can immediately train the binoculars on the track, and, if the satellite happens to be running early on prediction, you will catch it instead of letting it slip by while you are still searching for the right place to look. Looking near familiar stars is not always wise, however: for example, the satellite may be at low elevation or about to go into eclipse, or at an unfavourable phase angle (ie between you and the Sun). Often it is best to observe when the satellite is at its highest elevation and probably brightest. Then the track appears nearly horizontal, so that you can sweep the binoculars back and forth along the track without having to check which stars it passes.

You are now ready to go out and observe. This is where I find myself in difficulty, because nearly all observers use different techniques for observing and, by describing one method, I may give the wrong impression that this is the 'correct' method. So let me say again that there are as many ways of observing as there are observers. There are no rigid rules: if you achieve good results with your method, then it's good — for you. I shall merely describe how I usually observe and then mention some of the possible variations later.

Three minutes before the satellite is due, or a little earlier if I suspect that the prediction is in error or the satellite is very faint, I put on an overcoat with a hood, place two different stopwatches and a pencil in the right-hand pocket, a torch in the left pocket, hang 11 × 80 binoculars round my neck, tuck Norton's Star Atlas under my arm, switch off any objectionable lights, grab my deck-chair and venture forth. Most observers go into their back garden, but I usually observe from a (railed) flat roof: here the visibility is better, but the garden is much darker, and better for very faint satellites. If no rain is falling, I set the deck-chair at the lowest notch, sit carefully down, and glance at the star atlas with the aid of the torch, to refresh my memory of the predicted track among the stars. (A card with the night's predictions marks the right page of the atlas.)

Then I look up to find the chosen stars, and this is a critical moment: I may not be able to see them! Perhaps a cloud-bank has silently arrived — a serious hazard in Britain. Perhaps it is not quite dark. Perhaps there is a full moon and haze. Or perhaps someone has switched on a light in the house. If I can see the chosen stars, I immediately direct the binoculars towards them, at the same time sliding down to a comfortable reclining position — the deck-chair has a slippery plastic fabric. Observers who wear glasses (and many do) are advised to hang them on a cord, like the binoculars. (Some observers, including myself, ignore this good advice and hold their glasses with their teeth, only to receive a well-earned rebuke from their optician, 'I wish you wouldn't chew the ear-grip'.)

As soon as I have the chosen stars in view, I sweep the binoculars along the expected track, to try to catch the satellite if it is early, and then bring the field of view back to the area chosen for observing and, holding the binoculars with the left hand, take out the first stopwatch. Ideally an observer needs three hands, but I have to make do with two, and I usually hold the binoculars with the left hand and support them with the palm of the right hand, clutching a stopwatch with the fingers, Fig 35.

If the satellite has not already arrived, I swing the binoculars about 15° back and forth along the expected track and try to decide on the best reference stars to use. Choosing beforehand is useful, but you must always be ready to scrap your choice if the satellite is a little off the predicted track when it appears.

If the satellite has still not arrived, you should keep the binoculars steady, and wait for it. Good advice; but few observers take it, and most would admit that they sweep their binoculars to and fro along the track. You can find plenty of excuses for not

Fig 35 *Visual observer in restful action. Note the mobile observatory (deck-chair), the screen of evergreen trees (to block lights and cold winds), and the need for three hands.*

keeping still. It is less tiring for arms and eyes to sweep than to hold the binoculars steady. You can often spot the satellite before it reaches the chosen stars and give yourself time to make a better choice of reference stars. The satellite may be faint as it passes the chosen stars and brighter elsewhere, so that you have a better chance of seeing it if you sweep. But if you wave the binoculars around too wildly, you will probably miss the satellite, and possibly lose sight of the chosen stars: so don't sweep like a maniac, but take it quietly.

Now 'the moment of truth' is imminent: or is it? The suspense is one of the fascinations of observing. Were the predictors false prophets? Has the satellite decided to become faint? Have you used predictions for the wrong day, or misread the time, or done something equally stupid? All these questions niggle you as you watch; your arms are aching, your eyes are tired, your hands may already be frozen; your depression deepens as each second ticks by. Then you spot the satellite at last, and feel again the thrill that Keats could never have known except in imagination:

> Then felt I like some watcher of the skies
> When a new planet swims into his ken.

71

But this is no time to be starry-eyed. You need all your powers of alertness and concentration to make three accurate observations of the satellite during the minute or two before it enters eclipse (or goes behind trees, clouds, the roof of the house, etc). Remember that one observation is infinitely better than none; two are better than one; three are rather better than two, and give you more confidence because you can check that they are roughly on a line, or a gentle curve, in the star atlas.

To make an observation, you take a pair of not-too-sparkling stars A and B, preferably not more than 1° apart, such that the line AB between them is nearly perpendicular to the satellite track, and estimate the position of the satellite as it crosses AB (Fig 30). At the same moment you start your stopwatch. If you have a split-action stopwatch, you stop the subsidiary hand when you make the second observation. For the third observation, you have to use another stopwatch. Finding three suitable pairs of stars is not easy, and you will sometimes need to use other patterns, as we shall see in the next chapter, which is devoted to the tricks and variations.

Having made the observations, I switch on the torch and write details of the observations in the star atlas before I forget. If the reference stars are marked in the atlas, I write '.6 up' (or whatever fraction it is) in pencil, with an arrow pointing between them. If I don't know the stars, I sketch their pattern in the margin of the atlas. The stellar magnitude of the satellite and its variations should also be noted before you forget.

The next step is to go in and stop the stopwatches against the Speaking Clock time signals. On hearing the first Speaking Clock announcement, I write down a time 30 seconds ahead. Then, if using a watch with a rotating hand, I note the tenths-of-a-second reading at each of the three pips 8, 9 and 10 seconds later, to decide the correct tenths-of-a-second reading, and check again at the three pips 18, 19 and 20 seconds later. Then I stop the watch at the time already written down, 30 seconds later, and read the watch, perhaps 2 minutes 36.3 seconds, the tenths-of-a-second reading (.3) being as determined earlier. (It should also be the time shown by the watch after being stopped, but if your hands are frozen, you may be clumsy in stopping it.) Subtracting this reading from the Speaking Clock time gives the time of the observation. The reading of the split-action hand is then added, to give the time of the second observation. After checking that both readings were correct, I return the stopwatch to zero. The result should be accurate to 0.1 second if the observation was cleanly and decisively timed.

If I have taken a third observation using stopwatch No 2, I write down two times, 30 and 60 seconds ahead of the first Speaking Clock announcement, and stop watch No 2 against the later signal. (After long practice you can omit one of the timing checks and write down times 20 seconds and 40 seconds ahead of the first announcement.)

If you use a digital stopwatch with 'lap time' facility, the 'lap time' is equivalent to the split-action hand reading of a dial watch. The time of the second observation, as given by the 'lap time', should be read off first, written down and erased from the display. Then the 'lap time' switch should be pressed and erased at several time signals, to obtain a consistent tenths-of-a-second (or hundredths-of-a-second) reading. Finally, you stop the main watch against the Speaking Clock, using the tenths- (or hundredths-) of-a-second reading already established.

If all has gone well, it is now six minutes since I put on my coat to go and observe, and I have three timed observations and sketches of their positions. All that remains is to work out their RA and Dec. Norton's Star Atlas is not suitable if your observations have an accuracy better than 0.2°, because the scales are curved and so the accuracy is often poor. To find the RA and Dec accurately, you should use the superb star atlases of A. Bečvář, the *Atlas Borealis* covering declinations from 30°N to 90°N, the *Atlas Eclipticalis* for declinations between 30°S and 30°N, and, for southern-hemisphere observers, the *Atlas Australis* (see 'Further Reading'). These atlases include all stars ever likely to be visible with 7 × 50 binoculars, and have transparent linear grids for reading off position. You also need a transparent metric ruler. The declination scale in the Bečvář atlases, 2 cm to 1°, is very convenient because 0.05° is 1 mm, which is the marking on a normal ruler, and the average error in measurement is about 0.01°, as can be confirmed by reading off star positions from the atlases and checking against a star catalogue. Finding the RA and Dec of three observations takes about five minutes if all goes well, that is, if you find the reference stars quickly and they are not on different pages of the atlas.

Finding the RA and Dec of the observation in the atlas is straightforward, but needs the utmost care and attention to detail. The centres of the circles representing the reference stars should be carefully marked with a sharp pencil, and the distance between them (say 4.6 mm) carefully measured and multiplied by the fractional distance of the satellite (say $\frac{3}{10}$ from star A). The satellite's position should then be carefully marked by a point, with a circle round it, at the appropriate distance (1.4 mm) from star A. The transparent plastic grid should then be carefully held in place and the metric ruler used for the final measurement. The grid is intended to fit the underlying page exactly, but changes in temperature and humidity produce discrepancies, and you must automatically test for these and try to allow for any mismatch between the grid and the page. Occasionally, especially in the middle panel of the page in the *Atlas Eclipticalis*, this may be a limitation on your accuracy. If in doubt, re-measure and see if the second measurement agrees with the first.

If other satellites have to be observed, the measurements in the atlas can be done later, but I find it best to measure while the memory is fresh.

After some practice, the complete performance — going out and making three observations and recording them — will, if fortune smiles on you, take about twelve to fifteen minutes. The time taken over the predictions can be as little as 1 minute if you have 'look angles' at the ready, but much longer if you have to work out 'do-it-yourself' predictions.

This completes my account of how a satellite may be observed, but please remember that it is merely one possible method, and many variations are possible. These will figure in the next chapter, which is devoted to refinements and varieties in visual observing.

How to start

To conclude this chapter, I ought to give some guidance to anyone who is keen to start observing. The information is correct at the time of writing (1982), but is obviously liable to become outdated as the years pass.

A new observer in Britain or Western Europe would be well advised to join the British Astronomical Association (see page 176 for details). The BAA has an Artificial Satellites Section which, with the aid of a grant from the Science and Engineering Research Council, has sent simplified predictions to new observers. When observers become skilful enough, they are provided with the full predictions, either of the geographical-track type or 'look angles', by the British satellite prediction centre at the Earth Satellite Research Unit, University of Aston, St Peter's College, Saltley, Birmingham, which also sends predictions to many overseas observers who have proved their worth.

As well as receiving predictions, serious observers in Britain are also supplied with stopwatches, binoculars (usually 7 × 50) and the Bečvář star atlases. A stock is held at the Royal Society, London, and the equipment is issued on the recommendation of the Optical Tracking Subcommittee of the British National Committee on Space Research, which meets regularly at the Royal Society to coordinate visual observing activities.

Although I have recommended joining the BAA, this is not essential for UK observers and may be impracticable for observers in other countries. If, after reading this chapter, you think you can develop the necessary skills on your own, and if you go ahead and do so, you should send in a long list of your observations to the prediction centre, describing your methods and expressing your enthusiasm. You should find that the prediction centre will then supply you with predictions. But you must produce results: promises are not enough.

In the USSR and most of the eastern European countries, the Astronomical Council of the USSR Academy of Sciences acts as the centre for predicting and organizing satellite visual observing. There is a good exchange of observations between the USSR and UK, and the predictions sent out from the British prediction centre go to observers in eastern Europe, in Finland, Hungary, Romania, Poland and Czechoslovakia, as well as to the USA, South Africa, Australia, New Zealand, Japan and to western Europe.

In countries other than Britain and USSR-plus-eastern-Europe, there are at present no 'official' prediction services specifically for visual observers. However the Goddard Space Flight Center in the USA kindly provides the orbital elements on all satellites determined by the US Air Force, and many observers take advantage of this service. (The address is Code 512, GSFC, Greenbelt, Md 20771.) The British prediction centre also relies on these USAF orbits for most of its routine predictions, and so do numerous 'private' prediction services round the world. These usually consist of a keen observer or a professional group who supply predictions to local enthusiasts. They change as time passes, so it is not appropriate to give details.

By one means or another, most of those who are keen to observe do find a way of acquiring predictions. Perhaps in a perfect world anyone who wished to observe would be able to obtain 'look angles'. But such a service would be costly and impracticable in today's far-from-perfect world.

6
Visual Arts

But as each mass the solar ray reflects,
The eye's clear glass the transient beams collects.

Erasmus Darwin, *The Temple of Nature* (1803)

Chapter 5 was a breathless gallop through the essentials of observing, rather like the Charge of the Light Brigade — 'theirs not to reason why'. Many of the unasked questions crop up in this chapter and, if not answered, are at least discussed. Here I look at what might be called the culture of visual observing, the 'tricks of the trade' that superlative observers take for granted but the less skilful never think of.

Making a good prophet

'Methinks I am a prophet new inspired', gasped the expiring John of Gaunt. The prophets at the prediction centre may breathe in atoms from his last gasp, but they need more than his inspiration to succeed in forecasting the Sun's future activity. They often have to assume that 'as things have been, things will remain', and hope that the satellites themselves are intoning the prayer:

> God bless the Sun and his relations,
> And keep us in our proper stations.

Sometimes solar-terrestrial relations *are* smooth, and the satellites *do* remain in their proper stations; but storms in the Sun can disturb the upper atmosphere and make the satellites run erratically. This is just the time when observations are most needed because, by determining the satellite's orbit when it is unpunctual, we can find the unusual forces which must have been acting on it. Searching for ill-predicted satellites is one of the many arts of observing.

The satellite prediction centre at the University of Aston maintains an ansaphone correction service (present telephone number 021 327 2925), but there are always a few satellites for which the corrections are unknown because of the lack of recent

76

observations or orbital elements; these are quite likely to be the important satellites, particularly those close to decay, which are very difficult to predict accurately.

The predictions are computed by taking the orbital elements of a satellite (inclination, orbital period, etc) at a particular date, assuming a value for the rate of decrease of period — perhaps the same value as the previous week, or perhaps different if there is a good reason. So, if we exclude 'misprint-type' errors, which completely ruin the predictions, the most likely source of error in prediction is a change in the actual rate of decrease of orbital period, caused by a change in air density — itself attributable to an unexpected burst of particles and radiation from the Sun.

If the predictors assume a wrong rate of decrease of orbital period, the error in the predicted time will increase as the *square* of the time after the date of the orbital elements used. So if you observe a satellite running 4 minutes earlier than predicted on Monday, when the predictions are (say) seven days old, and if Tuesday and Wednesday are cloudy, you would guess on Thursday, when the predictions are ten days old, that the satellite would be $4 \times (10/7)^2 = 8$ minutes early. This is the best rule if the satellite is running early. But if it is running 4 minutes *late* on Monday, you would be wise to assume that the error is increasing linearly with time, and that on Thursday the satellite would be $4 \times (10/7) = 6$ minutes late, while bearing in mind that it will more probably be 8 minutes. It is better to waste two minutes waiting than to risk missing your quarry, waiting even longer, and losing your labour.

But suppose you have no idea of the correction, and the ansaphone only gives other satellites. Those other satellites often provide the clue: for if the air density increases, nearly all satellites with perigee below 500 km will be running early, and, if so, yours is likely to be running early too. If all the others are late, yours probably will be. (If the perigee height of your satellite exceeds 500 km, your best bet is to assume that the predictions are correct.)

A reasonable searching-time is about 10 minutes, although you can search as long as you like if you are keen enough. If you think the satellite is early, you can search during the 10 minutes preceding the predicted time. If you have no clue, you can search from 5 minutes before to 5 minutes after.

If the satellite is expected to be visible to the naked eye, you should look with the naked eye, which remains unequalled as a detector of large satellites at night. The eye has an unrivalled field of view, an excellent swivelling system and a keen perception of movement, coupled with a fine-angle resolution which makes radar seem clumsy. Once you have seen the satellite with the naked eye, you can transfer to binoculars just before the satellite approaches a bright star, and choose reference stars from among the faint stars near by.

Often, however, the satellite is unlikely to be visible to the naked eye, and the search is made with binoculars. Although the timing of the predictions may be in error, the position of the satellite's orbital plane in space is nearly always well predicted, because it only changes by a few degrees per day and its rate of change should be accurately known at the prediction centre. In searching, therefore, you should assume that the position of the orbital plane in space is correctly predicted. If so, the satellite's track *over the Earth* is, because of the Earth's rotation, 1° further east in longitude if the satellite arrives 4 minutes earlier than predicted, 2° further east if it is 8 minutes early, and so on.

You should use these changes of longitude to calculate roughly the change in the satellite's track if it is 5 minutes early (or late), and as you search you should gradually (at the right time) move the search area from the expected 5-minute-early track to the expected 5-minute-late track. This requires finesse, since you cannot stop observing to look at a watch. Some observers rely on mysterious internal clocks, but most find it best to count the seconds.

Fig 33 is most useful in estimating the change in elevation angle due to a timing error. Remember that 1° of longitude is equivalent to 111 km at the equator, 85 km at latitude 40°, and 72 km at latitude 50°. So if a satellite 500 km high is predicted to pass at 66° elevation due east, it will be 200 km east of you, from Fig 33. If it arrives 4 minutes early, it will be 1° further east, that is, 272 km east if you are at latitude 50°, and its elevation will, from Fig 33, then be 59° instead of 66°.

To minimize the problems of searching, it is best to look for the satellite at the lowest elevation at which you can be confident of seeing it in the prevailing weather conditions. When a satellite is passing overhead going south-north at latitude 50° at a height of 200 km, a 5-minute error in time alters its maximum elevation by about 25°, as Fig 33 shows. But if you can observe it when it is 500 km away, preferably to the east, its predicted elevation of 19° will only be reduced by 4°, to 15°, if it is 5 minutes early; this is within the 5° field of your binoculars, and all you need do is to look a little lower at the start and a little higher at the end of the 10 minutes. If the satellite is due to come overhead, you can look for it 500 km away in the south: if it is 5 minutes early it will be at an azimuth about 10° further east than predicted.

If the satellite is faint, you have to look for it at high elevation; but there is one trick you can play if your latitude and the satellite's orbital inclination do not differ by more than about 5° (if it is at 200-400 km height) or 10° (if it is higher). The trick is to observe the satellite at apex, the point of maximum latitude, where a timing error has little effect on the track. Fig 36 shows the predicted track, and the track when 12 minutes late, of a satellite with an inclination of 49°. If its height is 500 km or more, it can be observed at apex at latitudes from Scotland to Spain (if it is bright enough), and the track among the stars will be very little altered by an error of up to 15 minutes in time.

Observing ill-predicted satellites overhead is not nearly so difficult if they are at a great height: on Earth, distance lends enchantment to the view, it is said; in space, distance lends enhancement to the field of view. For example, a polar satellite at a height of 4000 km, passing overhead at latitude 50°, changes its elevation by only

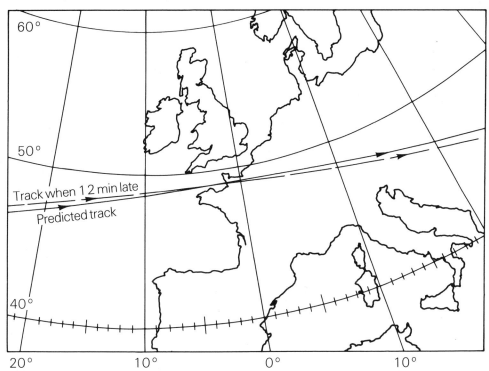

Fig 36 *If the predicted time of a satellite transit is in error, it is best to observe the satellite near apex (maximum latitude), because a timing error has little effect on the track near apex. The example shown is a 49° satellite with predicted apex longitude 0°, when it is on time or 12 minutes late.*

1.6° if it is 4 minutes late. Relative to the stars, the change is even smaller, about 1° and well within the field of view, because the stars themselves move by 0.6° in the same direction during the 4 minutes.

If the prediction centre is doing its job well, searching for ill-predicted satellites should rarely be necessary. But in practice the need does often seem to arise, particularly with newly-launched or decaying satellites. An observer struggling with a badly predicted satellite is like a yachter who has fallen into the sea: you shouldn't be there, but if you are, you need to know how to cope. Sharks are unlikely to disturb an observer, but rogue satellites can be very distracting. If you search for 10 minutes with binoculars in a direction where the shadow is not too high, you are very likely to see another satellite, and you need to decide almost instantly whether it is yours or not. The direction of travel is the best clue, and the angular speed the second best; but if the direction and speed are similar to your satellite's, you will have to follow the one you see and take observations of it. You can only decide afterwards, from

79

the track in the atlas, whether it was yours. If not, bad luck! It's too late now. Sometimes you can never be absolutely sure whether you have seen your satellite or another piece from the same launch, or another satellite in a similar orbit.

If you cannot obtain predictions, you can always observe in this random fashion. You just look through the binoculars until you see a satellite, and make observations. That is easy enough. The problem is to identify the satellite, which takes time and skill. And there is no guarantee that anyone will be prepared to devote their time to identifying your satellite.

Split-second timing

An observer has to record both the time and the position of a satellite, and simultaneity of hand and eye is a necessary talent for a good observer. Time and position are like the twin strands of electrical wiring: both are essential and they are intertwined; but they are also separate. Timing is the simpler, so I shall discuss it first.

'Reading maketh a full man; conference a ready man and writing an exact man', Francis Bacon tells us. The third of these maxims is the weakest, because many writers are inexact (just think of journalists), and we might offer Bacon an amendment (and remove his sex bias): 'observing maketh an exact person'. The good satellite observer should always be desperately conscious that visual observations are less accurate than photography, and should strive unceasingly for better accuracy in timing and position. Split-second timing is not good enough: the error needs to be less than a tenth of a second, if humanly possible.

There are limits, of course. A film run at 24 frames per second appears to the human eye as a continuous 'movie' rather than a series of jerks, because the response time of the eye is greater than $\frac{1}{24}$ second, that is, 0.04 second. The average response time of the eye is about 0.05 second in the young and healthy, and it seems unlikely that timing much better than this is ever regularly achieved by satellite observers, even if hand and eye are perfectly coordinated.

Though it is necessary to begin with this disclaimer, all the indications from practical tests are that most observers (and non-observers) can come very close to the best possible accuracy, if conditions are good. One such test was conducted at a meeting of visual observers at the Royal Society, London in 1968. All the participants were provided with stopwatches, and when a spot of light moving across a screen in a darkened room passed a fixed pair of 'stars', the 'observers' started their stopwatches, and then stopped them soon afterwards against time pips. The watches were read and the experiment repeated many times. The average scatter of the readings about the mean was 0.05 second. Other tests have given similar results. The implication is that an observer with an accurate watch who really tries hard, and observes a fast-moving satellite passing a pair of stars perpendicular to its track, can often achieve an accuracy of about 0.05 second; and most observers should be able to achieve 0.1 second, even in less favourable circumstances — eg with freezing winds, a watch that only reads to 0.1 second, or an oblique pair of stars.

If the satellite is slow-moving, the timing is inevitably poorer. The limit of resolution with hand-held 7 × 50 binoculars is about 0.01° — that is, two stars 0.01° apart can only just be seen separately. So a satellite moving at 0.01° per second could not be timed with an accuracy better than 1 second. Fortunately, you will never see satellites moving as slowly as this, unless you deliberately try for those at distances greater than 7000 km. A likely minimum is about 0.05° per second, and the timing error is then likely to be at least 0.2 second.

Apart from very slow satellites, what other gremlins may spoil your attempted 0.1-second accuracy? The first suspect is the stopwatch itself. Obviously you must check against time signals that it is consistently accurate to 0.1 second over the normal time interval of use — 5 minutes or so. If not, you must make a correction, or get a new watch. If it is a dial watch with a winder, you should check whether it goes faster when fully wound: if so, you should wind it fully before use, and make allowance for any consistent error. You should also test the effects of low temperatures, and make checks at several points on the dial, in case the hand is not centrally mounted. The delay in the starting and stopping of the watch after the knob is pressed may sometimes be troublesome: but such delays do not matter if they are the same at starting and stopping; again this can be checked against time signals. If you have a quartz digital stopwatch, its accuracy is likely to be much better than yours, but it is still worth testing. You can also test your own consistency by starting the watch at random and stopping it with the 'lap-time' switch at (say) 12 Speaking-Clock time signals, every ten seconds. You may well be surprised at your own accuracy: the scatter about the average value is often as low as 0.03 second, and you can usually tell whether you are early or late before you look at the reading displayed.

The next possible culprit is the time signals. But the time signals used by observers, such as the Post Office Speaking Clock or radio time signals in Britain, the WWV radio signals in the USA, or DIZ in Europe, normally have errors less than 0.05 second. So, on the basis of past behaviour, this suspect is 'not guilty'. But in such an uncertain world as ours, the possibility of the time signals becoming less accurate cannot be ruled out.

The third possible offender is your personal error in assessing the time when the satellite crosses the line between two stars. Ideally, as the practical tests showed, this can be as little as 0.05 second. But the error will obviously be larger if the stars are far apart and the 'line' between them becomes ill defined. So, for good timing as well as accurate positioning, you need to use stars less than $\frac{1}{2}$° apart, if possible.

A fourth possible culprit is your personal 'reaction time', the time delay before your hand (or foot, if you are driving a car) reacts to an event. This delay is important in, for example, timing the disappearance of stars behind the Moon in occultations.

But it is usually irrelevant in satellite observing, because you can anticipate both the movement of the satellite towards the reference stars and the final time pip in a series of three or six. However, if the satellite is only visible in flashes and you start the stopwatch when you see it flash (also memorizing the pattern it forms with nearby stars), you should subtract 0.2 second from the time you obtain, and quote an error of 0.2 second rather than 0.1. If you know that your reaction time is 0.4 second, you should of course subtract that. But when the satellite is continuously visible, this suspect must also be cleared of all blame.

A fifth possible source of error in timing arises from the stopping of the watch against the time signal. This error should rarely occur if you take the precautions already suggested, and decide on the correct tenths-of-a-second reading by estimating at one or more previous sets of time pips either from the moving hand or by using the split-action hand (or 'lap-time' reading).

Good observers avoid the possible dangers and split their seconds most reliably, to the relief of the analysts who use their observations.

Timing has always been rather a thorn in the flesh of visual observers, and some subtle methods of avoiding the timing error have been proposed. Most are basically similar to that suggested by V.I. Belenko. The observer uses a telescope fitted with a shutter, and also has a chronograph to register exactly the moment when the shutter closes. The observer operates the shutter when the satellite is passing suitable stars and memorizes the satellite's position at the moment when the scene is blotted out by the shutter. The timing error is very small if the chronograph is accurate, but the directional error may be increased. So the method is probably best for fast-moving low satellites, if you can obtain the extra equipment. There remains some doubt, however, whether these sophisticated methods eliminate the timing error or merely transfer it to the directional error. The methods are more successful with theodolite observations than with visual observations relative to the stars.

Many observers prefer to do their timing by starting their stopwatch(es) against a time signal, taking the running watches outside and stopping them at the times of the observations. With this procedure the correct starting of the watch should be checked by reading it at several subsequent time signals before going out to observe. One advantage of this method, if you can manage such superb organization, is that the watches can be read while you are outside, returned to zero, and then used again with the normal timing procedure if a second satellite is imminent.

Observers fortunately do not have to worry about the exact definition of time, but often ask, 'What are AT and UT?' So it is worth mentioning the three brands of time relevant to observing. First, there is Atomic Time (AT), in which the second is defined as a specific number of vibrations of a caesium atom. Most people assume that Atomic Time proceeds 'regularly', at a 'constant' rate: that may be true, or may not; but Atomic Time certainly proceeds more regularly than the other brands of time, particularly Universal Time 1, or UT1, which is tied to the rotation of the Earth, and increases by 86400 seconds every time the Earth rotates through exactly 360° relative to the 'mean Sun', or 360.98563° relative to the stars, if you don't like the concept of a 'mean Sun'. All satellite orbit calculations have to use UT1, because the angular rotation of the observing station about the Earth's axis, that is, its position in space, is determined by the time of the observation in UT1. But UT1 is non-uniform because the Earth is slowing down, at a variable rate which is linked with atmospheric winds and possibly earthquakes. So UT1 is inconvenient to use in scientific work that is not affected by the Earth's rotation. The third brand of time is 'coordinated Universal Time' or UTC, which proceeds at the same *rate* as Atomic Time but has one-second jumps (on 1 January, and if necessary, 1 July) to keep UTC as close as possible to UT1. If these 'leap seconds' were not introduced, the hours of day and night would, over thousands of years, gradually become interchanged: bright sunshine at midnight and darkness at noon would be very confusing.

The broadcast and telephone time signals are all in UTC, which is normal civil time, and all satellite observations should be timed in UTC. When orbits are determined from the observations, UTC will be converted to UT1, but that is the job of the orbit analyst, not the observer. No one knows beforehand the exact difference between UTC and UT1, and the value each day is estabished in retrospect by the Royal Greenwich Observatory at Herstmonceux, and sent to the satellite computing centres. The difference should be less than 1 second, unless the Earth changes its spin rate as a result of some disaster, such as unprecedented earthquakes or worldwide hurricanes.

Pin-point accuracy

Split-second timing is no good without pin-point directional accuracy, and that is not a literary metaphor, but an accurate description: your observation should ideally be so accurate that it needs a pin-prick rather than a small circle to represent it in the *Atlas Borealis* or *Eclipticalis*. You should strive to be as accurate as the time signals in timing, and as accurate as the *Atlas Borealis* (0.01°) in position. When you acquire such a sense of exactness that you feel you are adversely affected by the 0.01° error in reading the *Atlas Borealis*, you can use the *Smithsonian Star Catalog*.

But, before descending to detail, perhaps I may ask you to inject yourself with a strong dose of amazement that an observer with no more than a pair of binoculars can achieve such accuracy — better than the best performance of the largest multi-million-dollar tracking radars. Satellite observers could not ask for a better system for accurate directional measurement than is provided by the stars. They are point-sources of light; they remain in virtually the same position year after year; they differ in brightness and are disposed in distinctive patterns so that they are quite easy to identify with the aid of star maps — just imagine the problems if they were in a rectangular grid and of equal brightness! Although the brighter stars have been used for astro-navigation in the past, satellite observing is probably the first human activity to make full and regular use of all the fainter stars. This ironic fact has provoked many comments, mostly frivolous: 'at last astronomy has a human use'; 'will the fainter stars like being noticed, and reward us?'; 'when God made the heavens, He or She knew that people would observe satellites — that's what the stars are for'. More seriously, it does seem that satellite observers, together with comet-hunters and a few others, can claim to be the most complete of astronomers, because they personally use all the stars down to magnitude 9. Professional astronomers are mainly astrophysicists, who would not relish being asked to measure up, with a pencil and ruler, the positions of more than 10000 stars, as many satellite observers have done (including myself). Satellite observers may not know whether a star is a red dwarf, a blue giant or a yellow hole, but they certainly make good use of their heaven-sent system for measuring direction accurately.

And it is so simple too. You can achieve pin-point accuracy by following the method already described — choosing two close faint stars lying perpendicular to the satellite's track, and estimating the fractional distance between them as the satellite passes. The stars should not be more than 0.2° apart, and preferably within 0.1°. Achieving perfection is easier said than done, and in discussing the flaws that creep in, I shall begin with the imperfections caused by the brightness of the satellite and stars.

The most obvious problem is that the sky may be bright — perhaps the Sun has only just set, the Moon is bright or the sky is hazy. Faint stars cannot be seen, and the satellite is unlikely to pass two close bright stars. You cannot expect to make a perfect observation in these conditions. Using more powerful binoculars will help, but not much.

The second problem arises when the satellite is bright. Unless you are very lucky, the only reference stars close enough will be faint, and they may be lost in the glare as a bright satellite passes. This degrades your accuracy in judging distance and also the timing. The worst sets of observations I have ever handled when determining an orbit were for Skylab 1 rocket (1973-27B), which was the brightest satellite visible in 1973 and 1974. Many observers combat the problem by observing bright satellites at low elevation, where they are fainter. This solves one problem but creates another, because the position of a satellite can be 'fixed' more accurately if it is nearer. For example, a directional error of 0.01° on a satellite 1000 km away gives a positional error (perpendicular to the line of sight) of 200 metres; and so does a directional error of 0.05° on a satellite 200 km distant. To determine accurate orbits, it is the positional error that needs to be minimized; so the relative merits of low and high elevations are not always clear cut.

The two rules 'the nearer the better' and 'the fainter the better (provided you can see it)' lead to a conclusion which surprises many people: the most accurate positional observations of all are those made on very faint satellites when they pass overhead at heights near 200 km. If you ever observe such a satellite with an accuracy of 0.01°, you will be fixing its position cross-track correct to 40 metres. This is the ultimate in visual observing. The timing error should also be minimal for such a fast satellite, but even if your timing is accurate to 0.05 second, this still corresponds to an along-track error of 400 metres, since the satellite travels at about 8 km per second.

The relative brilliance of the reference stars can be a problem, because the fractional distance between them is more difficult to judge if one star is much brighter than the other. If you use a bright star with a faint companion as reference stars, your error is likely to be one tenth rather than one twentieth of the distance between. But an unequal pair may still be well worth using if they are much closer than any suitable equal pair. A slight difference in brightness is not important, and a pair of stars of magnitudes 7 and 8, or magnitudes 8 and 9 (if you can see them), will serve well.

Although a faint satellite can be observed more accurately than a bright one, you should beware of a faint satellite that varies rapidly in brightness and is invisible for half the time. You cannot achieve the best accuracy if the object is invisible as it passes the reference stars. Worse still, of course, the satellite may be too faint to be seen at all: that problem is discussed later in this chapter.

Even if you have no trouble with brightness, plenty of other imperfections can plague you. Perhaps you have begun to doubt your accuracy in measuring star positions in the *Atlas Borealis* or *Eclipticalis*: is humidity producing a mismatch between the paper of the map and the plastic overlay? Does the heat of your hand distort the grid on the overlay? If you are troubled by such doubts you should resort to a star catalogue, and the *Smithsonian Star Catalog* (see 'Further Reading') is recommended for this purpose. It gives all stars down to photographic magnitude 10 and, although the visual and photographic magnitudes sometimes differ appreciably, the *Smithsonian Star Catalog* includes all stars likely to be used as reference stars by observers with binoculars. If your reference star does not appear in catalogue or atlas, it may be a minor planet, a variable star, or a nova — it does happen, as when I tried to use Nova Delphini three days before it was discovered in 1967.

The right ascensions and declinations given in the *Smithsonian Star Catalog* and the *Atlas Borealis* and *Eclipticalis* are for epoch 1950.0: the position of the celestial pole and ♈ are taken to be as on 1 January 1950. The north pole position does not remain fixed relative to the stars, because it marks the point where the Earth's axis cuts the celestial sphere, and the axis of the Earth, like that of a spinning top, undergoes a steady precession, tracing out a circle on the celestial sphere once every 26000 years. The circle is centred on the pole of the ecliptic, which is roughly halfway between Polaris and Vega: the north pole is only temporarily near Polaris, and 13000 years ago it was near Vega. Observers do not have to worry about the movement of the pole; they merely state the epoch of the star map used. This will normally be 1950.0, but no doubt there will be a move to 2000.0 as that epoch approaches. The position of the pole shifted by 0.027° between 1950 and 1980, so the change is by no means negligible, and is allowed for when determining orbits from observations.

Some pages ago I remarked that the stars stay in *virtually* the same position year after year. Their positions are of course quite unaffected by the movement of the pole, which only affects our method of measuring their position — RA and Dec. However, the stars themselves *are* also moving relative to the Sun, at speeds of up to hundreds of kilometres per second, and this is known as their 'proper motion' (the apparent movements due to the Earth's daily rotation and 26000-year polar precession being regarded as 'improper', ie apparent rather than real). The angular proper motion of stars varies greatly, depending on their velocity and distance from the Sun. The proper motions in RA and Dec per annum are given in the *Smithsonian Star Catalog* and, if you want to make your observations as perfect as possible, you should, strictly, take account of this effect, from 1950 to the date of your observations. Only a few observers make these corrections, which are generally very small, especially for faint stars.

After this climb to the peaks of perfection, we now return to the practicalities of observing, and in particular the geometrical problems facing the observer.

The geometry of observing

Each satellite observation involves a geometrical judgment — or, if you wish to be really pedantic, an astrometrical judgment. You have to decide on the fractional distance between two stars, or the 'miss distance' from a star, or the shape and size of a triangle formed by two stars and the satellite. In discussing this subject, I am much indebted to an excellent paper by Geoffrey Kirby published in the *Quarterly Journal of the Royal Astronomical Society* in 1981, and I utilize several of his results.

We first need to know the resolution limit for hand-held binoculars, that is, the closest pair of double stars that can be separated. Kirby finds the resolution, in degrees, to be $0.06 \div$ magnification. Thus 7×50 binoculars would have a resolution of a little less than $0.01°$, while for 11×80 binoculars the figure would be about $0.005°$. These figures are an ultimate limit to observational accuracy for hand-held binoculars. Observers with firmly-mounted binoculars or telescopes have a lower ultimate limit.

I shall discuss four possible geometries for the observation, shown in Fig 37. Type (a) is the normal near-perpendicular passage between two stars. In type (b) the satellite passes outside the pair of stars. Type (c) is the 'miss distance' or appulse method. Type (d) uses a right-angle triangle. Many experiments have been made on 'observations' of type (a) and (b), by using a spot of light moving on a screen (the satellite) passing between two fixed spots of light (the reference stars). The most extensive tests were by R.H. Chambers in the 1960s. He found that for type (a) the accuracy is, as already indicated, generally better than $\frac{1}{20}$ of the distance AB between the stars; but the accuracy does vary with the distance AS of the satellite from the star A. Fig 38, adapted from Kirby's paper, summarizes the results for types (a) and

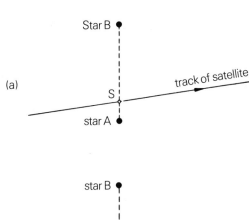

(a)

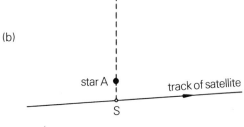

(b)

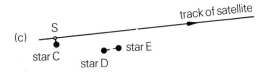

(c)

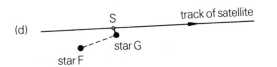

(d)

Fig 37 *Geometry of observation:*
(a) Near-perpendicular passage between two stars
(b) Satellite S passes outside the pair of stars
(c) 'Miss-distance' or 'appulse' method
(d) Right-angle triangle

(b) in the form of groups of values with error bars indicating the likely scatter (sd). The length AS is positive if S is between A and B, and negative if S is outside, as in type (b). The curve I have drawn, which may seem a little arbitrary, reflects my opinion (consistent with the results) that observations taken nearly midway between the reference stars are slightly more accurate than those one third of the way between. (The curve in Fig 38 could of course be continued beyond AS/AB = 0.5, and would be symmetrical about the line AS/AB = 0.5, the right-hand edge of the diagram.) The accuracy at AS/AB = 0, when the satellite track appears to pass through A, depends on the separation of the stars A and B. The value given, 0.02, is appropriate for an observer using 7 × 50 binoculars, with resolution 0.01° and

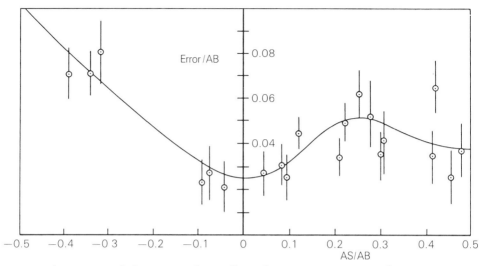

Fig 38 *The accuracy of observation of a satellite S between two stars A and B, as in Fig 37(a), expressed as a fraction of AB. If S is outside AB, as in Fig 37(b), AS is taken as being negative. The curve shows for example that if AS/AB = 0.3, an error of 0.05 AB would be expected.*

stars separated by $0.5°$ (since $0.02 \times 0.5° = 0.01°$), or using 11×80 binoculars and stars separated by $0.3°$.

Fig 38 confirms that observations of type (a), between stars A and B, are normally accurate to $\frac{1}{20}$ or 5% of the distance between them, and are quite often rather better than this. However, the observations of type (b), when the satellite passes outside the pair of stars, become less accurate as the distance increases, and should not be used by choice if the distance is more than $\frac{1}{2}$AB. By then the error is already 10% of AB and probably increases to about 15% when AS is equal to AB. Observations made outside a pair of stars are also likely to have larger errors in timing. So the rule is: always observe between a pair of stars if possible; if you must go outside, don't go more than half of the distance between them, unless they are a very close pair.

On a dark and clear night, observations of type (a) are recommended; but the 'miss-distance' or appulse method is needed if, for any reason, only bright stars can be seen, when there may be only solitary stars and no pairs visible near the track. This can happen if the sky is still bright at sunset, or with haze and a full moon, or if you have no binoculars. In the appulse method you take a mental photograph, as it were, of the distance CS in Fig 37(c), and immediately compare it with the distance between a pair of stars nearby, D and E, separated by a distance greater than CS. You then estimate the fraction CS/DE and measure up the distance DE in the star atlas, to give you CS. You then have to plot the predicted track in the atlas, and mark off the distance CS from the star C in a direction perpendicular to the predicted track.

The appulse method is more complex than the simple observation between two stars, and requires more practice to achieve a good accuracy. But it is a method that appeals to some observers, and since it is keenness that counts (as always), they will do well with it if they persevere. Kirby's studies show that a practised appulse observer can usually estimate the distance CS with an error of about 15%, unless CS is very small and the error is set by the resolution limit. So, with CS as 0.2°, for example, an observation would be accurate to 0.03°, if the direction is correctly assessed. Multiple observations are more difficult with the appulse method, because you have to remember and identify the reference star C and the star pair DE as well as remembering the fractional distance. Also you may lose the satellite while making sure of the identity of D and E, and be unable to make a second observation even if your memory is equal to it.

Although the fractional error is greater with the appulse method, there are more appulses than star pairs available for use as the satellite crosses the sky on a typical transit. Kirby's statistical analysis of the closest approaches and star-pair intersections, on a random 60° track across the sky, shows that by using the best star (or star pair) of magnitude 8 or brighter, an accuracy of about 0.01° should be attainable with either star-pair or appulse observations. The potentialities of the two methods are nearly equal. With the star-pair method, there is little to be gained by using binoculars with a magnification higher than about 10, or by using a star catalogue in preference to the Bečvář atlases. With the appulse method, however, a magnification of 20 gives an accuracy twice as good as a magnification of 7, and for these most accurate of observations a star catalogue is needed. If the night is dark and clear, and you have 11 × 80 binoculars or a telescope, you can use stars down to magnitude 9 and there may be two or three chances of achieving an accuracy of 0.01° on a typical satellite transit.

But do remember that 0.01° is an ideal. Few observers claim to achieve it and very few actually achieve it. Accuracies of 0.02° and 0.03° are more practicable and are regularly attained.

To return from the ideal to the practical, we have the fourth type of observation in Fig 37, the right-angle triangle. You time the satellite as it passes the stars F and G at the moment when the angle FGS is a right angle, and estimate the distance GS as a fraction of FG. This is rather like type (a), but the 90° turn leads to larger errors and you should assume an error of $\frac{1}{10}$ of FG (as compared to $\frac{1}{20}$ of AB in type (a)), provided GS is not greater than $\frac{1}{2}$FG. The errors arising through the angle FGS not being an exact right angle are usually about the same as the distance error.

Although the right-angle observation may seem intrinsically less accurate, it can be very valuable for impromptu use, especially with a distant and slow-moving satellite. You often find that a faint slow-moving satellite is about to travel close to a line of faint stars, UVWX say, as in Fig 39. You take observations at S_1, S_2 and S_3 with error $\frac{1}{10}$ UV or $\frac{1}{10}$ WX, that is, 0.03° if the stars are 0.3° apart. You can then plot

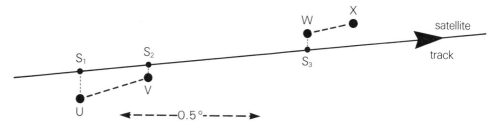

Fig 39 *As a satellite passes a line of stars UVWX, observations can be taken at S_1, S_2 and S_3. The distance US_1 is estimated as a fraction of UV, and WS_3 as a fraction of WX.*

the points in the star atlas and check whether they are in a straight line. If they are, you can be almost sure you have achieved the accuracy you hoped for, and it is good to have such a convincing configuration available so easily.

If you are lucky, the satellite in Fig 39, instead of missing all the stars, may appear to coincide with one of them. Such coincidences with stars are the most satisfying of observations, offering real pin-point accuracy, or so it might appear. If the satellite really coincides with the star, your observation would be as accurate as the resolution limit, $0.01°$ for 7 × 50 hand-held binoculars, and less for larger or fixed instruments. But I suspect that many 'coincidences' are really 'blendings', and I have never myself claimed an accuracy better than $0.02°$ on a coincidence. In a real coincidence, the satellite would eclipse the star, but only for a few milliseconds. If the satellite was faint and the star bright, there would be a diminution in brightness, but not for long enough to be seen with the eye.

Some observers like to look for pretty patterns, making an observation when a satellite forms an equilateral triangle with a pair of stars, a parallelogram with three stars, and so on. These special patterns cannot be recommended, because they rarely arise in practice and an observer who waits for them is tempted to believe that they occur more often than they really do.

Up to now I have assumed that the satellite is continuously visible. But many satellites are visible only in momentary flashes, and in observing them you will have to join the ranks of the pattern-fanciers. You cannot guarantee to observe such satellites accurately, although you may have a stroke of luck if the flash occurs directly between two stars. Otherwise you have to memorize the pattern made by the flash and the nearest stars, and try to pin it down by including at least three stars in your mental snapshot. It is best to wait until the satellite enters an area fairly rich in stars and to decide beforehand that you are going to use the next flash; then you must be ready to make your mental snapshot (and to press the stopwatch). Methods of remembering patterns seem to be very varied: some observers like to use lines, some use angles, some just 'visualise' and manage to retain their vision. In Fig 40, for example, you could say to yourself that the fourth flash is just outside the triangle

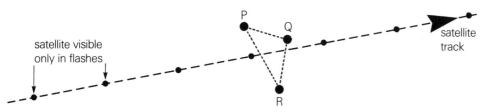

Fig 40 *Observing a flashing satellite by a pattern relative to the stars.*

PQR, about half way between P and R, and makes approximately an equilateral triangle with P and Q.

There are other aspects of the geometry of observing, and some of them arise later in this chapter under other headings.

How many fixes?

The question is simple enough, but the answer is not. What is best for one observer may not suit another. After you have made one observation it is obviously an advantage to make another of equal (or better) accuracy, provided you do not degrade the accuracy of the first (for example, by forgetting the reference stars). A third fix is of some further advantage in orbit determination and is valuable as a check on reliability, as explained in the next paragraph.

It is unwise to try to specify the 'value' of observations too exactly. Statisticians may mutter that n observations are $\sqrt{n}$ times more useful than one observation, and would say that three observations of accuracy $0.05°$ are equivalent to one of accuracy $0.03°$. But this is too naive, because an observation on an otherwise unobserved part of the orbit is much more valuable than an observation of equal accuracy on a well-observed part, and you are three times more likely to make one of these more valuable observations if you make three well-spaced observations rather than one. Also, when three observations are used in an orbit determination, timing errors are averaged out, and that is equivalent to 40% better accuracy on the $\sqrt{n}$ rule. Since there is always a shortage of observations for orbit determination, three observations help to improve the statistics of the calculation. Another advantage of taking three observations is that you define the track and can check that you are not observing the wrong satellite, always a danger if only one observation is made. You can also check your own reliability by seeing that the positions in the star atlas are on a straight line or a gentle curve. To summarize: there is no substitute for accuracy, but once you achieve accuracy, you will greatly improve your effectiveness as an observer by taking three observations instead of one (of the same accuracy).

'One observation only' is the general rule, however, for observers who use a telescope fixed in position and see the satellite for only a short time as it crosses the small field of view. They rightly concentrate on securing one very accurate observation, though more than one may be possible when the satellite is slow moving. The same preference is often found among observers who use mounted rather than hand-held binoculars. They too allow the satellite to pass through the field of view, rather than following it. There can be many other reasons for preferring to make only one observation: you may have only one single-action stopwatch; another satellite may be due, and you cannot wait to take three observations; and so on.

'Two or three observations' is the general rule for experienced observers with hand-held binoculars who have a split-action and an ordinary stopwatch. My favourite technique for observing is the 'three-in-a-line' method, taking three observations in quick succession, with gaps of about 10 seconds, or less if you can manipulate the stopwatches successfully. I look for a bright star near the track and then use faint stars near by, down to magnitude 8 or 9, as reference stars. The method works best for satellites that are faint (magnitude 6-9) and fairly distant (greater than 1000 km). I test the reliability by taking the time difference between the first and second observations, t_{12} say, and between the second and third, t_{23} say, and calculate a 'time ratio' $T = t_{12}/t_{23}$. Then I calculate similar ratios R from the three right ascensions, and D from the three declinations. If the observations are close enough together, the satellite (very nearly) follows a straight-line track at constant angular speed, so the three ratios T, R and D should be almost equal.

An example may clarify this 'ratio test'. Fig 41 shows my three observations of Cosmos 379, 1970-99A, on 9 October 1980 when it was at an elevation near 70° and moving from overhead down to the eastern horizon. The predicted track, as drawn in Norton's Star Atlas, passed just to the right of λ Andromedae, so I watched for the satellite there, in the hope that it would pass through the group of four stars to the right of λ Andromedae. Needless to say, the satellite chose not to do so, but made up for this perversity by coinciding with a star, marked as observation **1** and assessed as being accurate to 0.02°. The second observation, marked **2**, was at an equal distance beyond the two stars joined by the broken line: such extrapolation is not to be recommended, but it happened; my error estimate was 15% of the distance between the stars, that is 0.03°. For the third observation, marked **3**, the satellite again kindly coincided with a star, and the accuracy was again assessed as 0.02°. The three observations are nicely in a straight line, which is encouraging. The satellite was at a height of 4000 km: so it was faint (magnitude 7.5) and slow moving. The time between the first and second observations was 13.6 seconds, and between the second and third 20.0 seconds. The time ratio T is therefore 0.68, with an error of less than 1% if the timings are accurate to 0.1 second. So $T = 0.68 \pm 0.01$. The right ascensions at the first and second observations were 23 h 36.9 min and 23 h 41.8 min respectively, giving a difference of 4.9 minutes, which is possibly in error by 0.2 minutes (the equivalent in RA of 0.03°); the difference between the second and third RA was 6.9 minutes, so the RA ratio R is 4.9/6.9 or 0.71. Thus, allowing for the possible 4% error in the 4.9 minutes, we have $R = 0.71 \pm 0.03$. The declinations at the first and second observations were 45.44° and 44.96°, giving a difference of 0.48°,

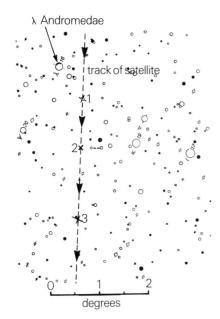

λ Andromedae

track of satellite

degrees

Fig 41 *Observations of 1970–99A on 9 October 1980, as plotted in the* Atlas Borealis. *The observations are marked 1, 2 and 3. The largest circles denote stars of magnitude 4, the smallest denote stars of magnitude 12.*

possibly in error by 0.03°: dividing by the second declination difference, which was 0.71°, gives the declination ratio D as 0.68 ± 0.04. The three ratios are therefore equal, to well within their error limits. This ratio test, plus checks that the track is parallel to that predicted, and that the angular rate of travel is as predicted, make it virtually certain that the observations are correct, in the absence of stupid errors such as writing down the wrong satellite, the wrong hour or the wrong day. Of course, the three ratios would not be exactly equal, because the angular rate of travel of the satellite does gradually change. The method should not be used for observations near the celestial pole, because the RA can change suddenly by up to twelve hours; nor on occasions when the declination is near a maximum, through the equality of T and R can then still be tested. Any observer worried about accuracy will find the 'ratio test' useful and often reassuring: I use it whenever I make three observations, as a guard against arithmetical errors.

The three-in-a-line method has the disadvantage that your observations are not widely spaced, and are therefore no more likely than a single observation to be on an unobserved stretch of the orbit. If the satellite is slow-moving, you can take three more observations ten minutes later, after recording the first three. And three more, ten minutes after that. With nine accurate observations at three widely spaced points, you would certainly be making a powerful contribution to the determination of the orbit.

Although I have emphasized the value of *accurate* observations, there are some observers who are not keen on accurate measuring-up, and find more satisfaction in making *numerous* observations. They might make up to twenty observations by a running commentary and a tape-recorder. Perhaps the simplest method is to record time pips a few minutes before and after the observations, and to describe the satellite's track: 'Satellite is now (sharp knock) passing about 0.1 degree below the top left-hand star in the square of Pegasus; now (another knock) three-tenths below two close sixth-magnitude stars perpendicular to the track; now (knock) six-tenths up between two mag 3 stars about 2° apart'; and so on. Afterwards you play back the tape and find the times of the knocks, with the aid of a stopwatch. Then comes the task of fitting the track in the star atlas to the description recorded. Inaccuracy is the main problem: the timing is less accurate because the tape speed is usually not quite constant, and the positions are estimated more hurriedly. I would not recommend this method, but it may appeal to some of those who know their way about the night sky.

A tape recorder, though far from ideal for timing, can sometimes be a valuable aid. If a satellite splits into a large number of fragments in line astern, a tape recorder is ideal for registering the times when they pass a line between two stars: 'First piece, now (sharp knock), five-tenths up, mag 8 ... second piece, now, six-tenths up, mag 5 to 7 fluctuating ... third piece, now ... ' If the fragments follow each other at intervals of only a few seconds, it is almost impossible to record their times with stopwatches.

Don't miss it

A naturalist trying to observe a shy species must stalk it cunningly; and the same goes for an observer in search of an elusive satellite, though the methods of stalking are different. As explained in Chapter 3, the brightness of a satellite depends chiefly on its size, distance and phase angle relative to the Sun, while the variations in brightness depend on its shape. A particular satellite, of specified size and shape, is likely to be brightest when you are facing away from the Sun as you look at it; also, of course, the nearer it is, the better. In practice, if the Sun is in the north-west, the satellite will be brightest in the south-east, and if it is faint, your best chance of seeing it is near maximum elevation in a direction between south and east. If it is a polar satellite which passes to the west of you, it will probably be easier to see in the south-west than in the north-west. If a satellite is low in the north, travelling from west to east, you have a better chance of seeing it in the north-east than in the north-west. Satellites are apt to enter eclipse in the evening when they are to the east, and may start to become fainter well before the eclipse point: so you should always observe 20° before eclipse, if possible.

'Observe with your back to the Sun' is generally a good rule, but you sometimes need to overrule it with another, 'observe with your back to the Moon'. During the week before full Moon, especially if the sky is hazy, observing near the Moon may be almost impossible, and you should observe from a place which is not in moonlight, even though that leads to a less favourable phase angle with the Sun.

A satellite that varies in brightness from 'easily visible' to invisible (as often happens with rockets) may go right across your field of view while invisible, if its rotation is fairly slow. You have a better chance of seeing it when it is lower in the sky, and takes longer to cross the field of view. Rotating cylindrical satellites may also sometimes break the rule of 'back to the Sun' by glinting most brightly when they are in nearly the same direction as the Sun. But this cannot be relied upon.

You also need perseverance in searching for satellites. Tennyson might almost have had the satellite observer in mind when he wrote: 'To strive, to seek, to find and not to yield'. You always strive for accuracy, you seek the satellite, you find it (if you are lucky) and never yield to despair if you fail. In observing, as in most walks of life, comfort promotes efficiency, and you will search more perseveringly if you are comfortable. For me, comfort is a deck-chair set at the lowest notch. But ideas of comfort differ, and some observers adopt postures that would be agony for me.

Having taken up your comfortable position, how should you search for faint satellites? The obvious technique is to direct the centre of the field of view of the binoculars or telescope towards a point on the track where all the nearby stars are faint, so as to set a standard for the satellite. You should keep the field of view fixed and let your eye roam over the field, upwards, downwards, sideways, or in circles. Staring fixedly at the centre of the field for more than about half a minute seems to be inefficient. Movement keeps the eye alert and prepared for the entry of the fast-moving satellite: otherwise its injection into the static picture can be rather traumatic, to the detriment of accuracy.

If you know the satellite varies in brightness and can be seen if caught at the right moment, you can try sweeping the binoculars along about $10°$ or $20°$ of the expected track, preferably at a speed slightly greater than the satellite's. You will catch it on the backward swing if it decides to be visible then; if not, you have a sporting chance on the forward swing, because it may happen to be within the field of view, though invisible, and its very slow movement across the field increases the chance of brightening.

If you think you are missing the satellite not because it is faint, but because it is off the predicted track, a backward zigzag scan (Fig 42) is a useful way of increasing the field of view from $5°$ to about $8°$. This situation can arise if you decide that the satellite is probably running late (because you fail to see it at the proper time). You then have to guess how much the track will shift, and the zigzag scan is your best ploy.

If the satellite is slow moving and you fail to see it at the first attempt, you can try again ten minutes later, further along the track, where you may be more fortunate. The best way of seeing a faint star is to look not straight at it but a little to the side, using peripheral rather than central (foveal) vision. This technique, known as 'aver-

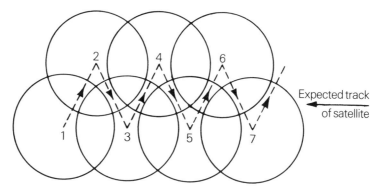

Fig 42 *Zigzag scan, to increase the effective field of view. The circles indicate the field of view of the binoculars at successive seconds of time.*

ted vision', is not usually needed for a moving object like a satellite, which is bound to appear first in the peripheral vision and will keep on reappearing in the peripheral vision if lost in the foveal. However, averted vision can be useful if the satellite is very faint and slow moving.

No one can see well on first going from a lighted room into the dark, and maximum sensitivity of night vision only comes after half an hour or so in the dark. In bright light you see with the cones of your retina, and when completely dark adapted, you see only with the rods. In looking for a moving light, such as a satellite, you should retain a good deal of the accurate cone-vision, and not reach the extreme state of dark adaptation when only the rods are in action. Maximum sensitivity of cone-vision is reached in about 5 minutes, and most of the improvement occurs in the first minute after going out. When searching for a very faint satellite you may benefit by going out 5 minutes beforehand to become dark adapted; but for normal satellites you need not bother to allow any time for dark adaptation. If you search for more than 5 minutes, eye fatigue outweighs any improvement in sensitivity.

This is all very well on a 'night of cloudless climes and starry skies', but what if the sky is cloudy, as so often in Britain? Some observers just give up, but perseverance is often rewarded. If the clouds are still and broken, you can look in the gaps, and the main problem is identifying the stars you see there: this is when you gain by knowing the stars well. If the cloud is high cirrus, you can sometimes see through it with binoculars, even if no stars are visible to the naked eye. In these conditions, moving satellites are much more easily seen than stars of the same brightness, and again the chief problem is to find and identify suitable reference stars.

If the clouds are broken and fast moving, observing becomes more of a sport than a science. Satellites have a notorious tendency to travel along behind a moving cloud. So if the wind is west-to-east you will probably do better with polar satellites than with those travelling west-to-east. Some observers like to follow gaps in the clouds, but this can be confusing because the eye has a tendency to treat the clouds as fixed and the stars as moving, so creating the illusion that squadrons of satellites are in transit. I prefer to select bright stars near the track, wait until they appear in a gap, and move to a place where they are in line with some vertical object, such as a television aerial, a solitary tree, or the post of a fence. You should also try to arrange that obstacles on the ground cut off vision of the sky (or clouds) at more than 10° below the expected track, as shown in Fig 43, so that you are not tempted to look low just because the sky happens to be clear there. You can then sweep the binoculars to and fro along the satellite's track, as defined roughly by the bright stars or the top of the obstacle.

Much more could be said about observing in cloudy weather. But I do not wish to give the topic too much emphasis: although more sporting than normal observing,

Fig 43 *When the sky is partly obscured by fast-moving clouds, it is an advantage to observe from a spot where some obstacle cuts off vision at 10° below the satellite's track and a vertical feature marks the most suitable reference stars. A fence with posts, as shown here, is ideal.*

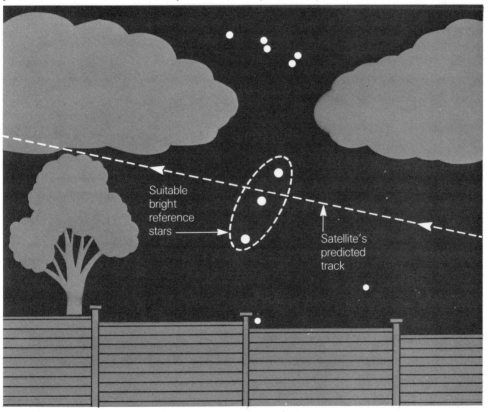

it can also be very frustrating, because you can look for several satellites and finish up with nothing to show for your labour.

Optical aids

The bare minimum for observing is the naked eye, that unrivalled detector of bright satellites. Good eyesight is obviously an advantage, but the wearing of spectacles is only a very minor disadvantage: about 40% of the most active observers wear glasses, including Russell Eberst, the world's leading visual observer.

Choosing an instrument for observing is very much a matter of personal preference. As already mentioned in Chapter 5, it is as well to have an aperture of at least 50 mm, a field of view near 5° in diameter, and an exit pupil of about 7 mm. A new observer is recommended to start with 7 × 50 binoculars and to progress to larger binoculars, such as 11 × 80, when the need arises and the money is available.

A wide field of view has many advantages. You will not be so likely to miss satellites that are badly predicted. You can find your way around the sky better. You can cover more sky when making speculative sweeps. You can follow the satellite across the sky more easily. You are more likely to see it if it is only visible in flashes.

Being able to track the satellite easily across the sky also has several advantages. You may see it first at an unexpected point of the field, barren of reference stars, and then you need to follow it until suitable stars appear. You usually need to track the satellite if you are going to make three observations, and to achieve the best accuracy, you may need to track it for perhaps 45° until it passes very close pairs of reference stars. You may wish to follow it to eclipse or to check that it is not an aircraft.

If you are already skilled in using a telescope, you may choose to observe satellites with a telescope; but telescopes cannot be recommended for new observers, because they are much more difficult to manage than binoculars. Despite these cautionary words, an observer with a large telescope has one great advantage over the entire binocular brigade — the ability to see faint satellites. Gordon Taylor of the Royal Greenwich Observatory has for many years used a telescope of 50 cm aperture to make accurate observations of satellites down to magnitude 14. His field of view is less than $\frac{1}{2}°$ so the predictions need to be accurate. In recent years he has succeeded in observing a number of communications satellites in synchronous orbits. Observers who wish to follow in his footsteps should start with a smaller telescope and develop their skills over the years. Observers with large telescopes can make a valuable contribution by observing satellites invisible to others, for example, by observing at apogee satellites that the binocular brigade can only see near perigee. Telescopic observations are also potentially the most accurate, because the faintest of recorded stars can be used for reference.

Telescopes firmly mounted may also allow better accuracy than hand-held binoculars because there is less wobble. This is an advantage, but probably not a very important one; few observers bother to make mountings for their binoculars, and most are content with hand-held binoculars. Mounted binoculars do offer some advantages, however, by reducing wobble and arm fatigue, and it is probably significant that Stanley Milbourn, one of the few observers to favour mounted binoculars, is perhaps the most accurate of binocular observers, and the only one regularly to attain an accuracy of 0.01°.

Non-observers are often surprised that quite heavy 11 × 80 binoculars can be held steady enough. The jitter is reduced by holding the binoculars at their centre of gravity, and a slight wobble causes little trouble, if you manage to hold the field of view steady for a few seconds before the observation.

'Whatever is most comfortable, is best', to misquote Pope, seems to be the maxim when choosing an instrument for observing. Looking with two eyes rather than one is an urge deep-rooted in human beings and animals, and most people find it uncomfortable to screw up an eye for long; and when they do, they find it difficult to unscrew, so that their features may become set askew (though it is possible to keep both eyes open or wear a patch over the unused one). We might summarize by saying that most people will choose binoculars; but skilled telescopic observers, and observers with one defective eye, may prefer a telescope. Whatever they choose they will probably grow fond of, because, to quote Pope correctly this time,

> To observations which ourselves we make,
> We grow more partial for the observer's sake.

So don't let an enthusiast mislead you into using an instrument that doesn't suit you. You know best what you like.

The observing site

Many people imagine a satellite observing station as having a view down to the horizon in all directions. Like Samuel Butler in *Hudibras*, they expect to find an observer living

> In mansion, prudently contrived,
> Where neither tree nor house could bar
> The free detection of a star.

On the contrary, an observer living near a town needs an area where the horizon is entirely screened up to about 20° or 30° elevation, either by dark buildings or dense evergreen trees. These barriers block the light round the horizon, which is inevitable near towns, and provide a dark area from which high-elevation satellites can be observed. The barriers also help to block moonlight and cold winds. From this dark area you watch for normal satellites, which should be observed at elevations above 30°. If no natural dark area exists, you can always build yourself a shelter with canvas or wooden screens. But since plenty of satellites can be seen, even in bad conditions, your rule should be, 'if you want to observe, observe'. (If you want to build shelters, do so by all means: but I doubt whether observing is your vocation.)

Although the dark area is the most important requirement, it is useful to have vantage points where you can see down to the horizon in one direction or another — perhaps a bedroom window or a flat roof. Failing this, you can sprint down the road to a better place: you can go roving 100 metres from your declared position, or even 200 metres if the observation is not of high accuracy; but you should 'go no more a-roving, so late into the night', as Byron said.

Byron also reminds us that 'the Moon is still as bright', and many satellite observers have been known to swear at it. In the evening, moonlight is usually most troublesome in the week before full Moon, when the Moon is to the south or southeast. So if you must live in a town, you should (if you have any choice) select the west or north-west outskirts and use the house to block both the Moon and the town lights. If the prevailing wind is south-westerly, as in Britain, the motto 'west is best' applies with greater force, because smoke from the town is less likely to drift overhead and degrade seeing.

In Britain a fairly high situation is probably better than a valley, both for seeing down to the horizon and for avoiding low mists. But the site should not be too exposed: otherwise winter winds may bring tears to your eyes, and shake the binoculars or telescope, usually with disastrous effects on the observations. Wearing a coat with a hood is strongly advised: the hood keeps out light from the Moon or the neighbour's windows, and also the cold winds. 'Observe with your back to the wind' is another maxim worth remembering if there is a howling gale, unless it leaves you looking towards a bright Moon!

Severe cold, even without wind, prevents long watches and has other ill effects. If your fingers are frozen, you will be clumsy with the stopwatch, and therefore less accurate, and you may even drop it. To avoid these hazards you can wear a glove with one finger-end cut off, or only remain out of doors for a short time. When a satellite is well predicted, I sometimes go out only about 30 seconds beforehand. A bright satellite at an elevation of less than 30° can often be observed from indoors through a window — if the window is clean! Observing from a reclining armchair indoors may not seem so virtuous, but it can be effective. And you can always say that tears in your eyes would prevent good observations if you went out into the cold.

Using a theodolite

So far I have been discussing visual observations relative to the star background, and this is the recommended method. But personal preference is paramount, and some observers prefer to use a theodolite, which does not need the stars at all, except for checking the accuracy. At its simplest, a theodolite is a small telescope mounted on a stand, perhaps a tripod. The mounting allows the telescope to be rotated in the horizontal plane over a wide arc, to look north or east, or in any other direction; and also to be elevated to a fairly high angle. There are scales to indicate the azimuth and elevation angles.

The observer follows the satellite with the telescope and stops the movement of the telescope just as the satellite is about to pass the intersection of the cross-wires in the centre of the field of view, at the same time starting a stopwatch. By reading off the azimuth and elevation scales, and obtaining the time in the usual way, you have an observation of the satellite in the form of time, azimuth and elevation. With practice, more than one observation can be made.

The accuracy of theodolite observations depends greatly on the quality of the instrument and the mounting as well as on the skill and assiduity of the observer. Great care has to be taken over the levelling of the instrument, and a concrete emplacement is advisable. The accuracy should be checked by taking readings on known stars — their azimuth and elevation have to be calculated. But if the emplacement, the instrument and the calibrations are satisfactory, a good theodolite observer can do as well as a star-background observer.

This simple procedure can be improved in various ways, and several professional observing stations use semi-automatic theodolites of high quality. With these instruments the observer follows the satellite, keeping it as close as possible to the centre of the field of view, as defined by the cross-wires, and presses a button when the satellite is central in the field. The readings of azimuth, elevation and time at this moment are then printed out automatically, and further observations can be made. Although their accuracy is no better than that of the star-background observer, such theodolites painlessly produce twenty or thirty observations per transit, a feat that no star-background observer could rival.

The observatory at Jokioinen in Finland has been outstanding for semi-automatic theodolite observations. Their work began in 1958 and their observers have made more than 30000 observations, which have been used in innumerable orbit determinations and have been particularly valuable because of the reliability and the high latitude of the station (62°N).

Steadies and flashers

So far I have been concentrating on accurate positional observations. But some observers detest measuring up distances in star atlases, and prefer a quite different form of satellite observing: they record the variations in the brightness of a satellite as it crosses the sky. The most assiduous such observer in recent years has been Horst Köhnke in Germany, but many other observers make positional observations and also record variations in brightness — indeed, this is regarded as part of the repertoire of a skilled observer.

Spherical satellites usually remain fairly steady in brightness as they cross the sky, but accurate estimates of their stellar magnitude are worth making as a check on the standard magnitude being used by the prediction centre, or, if the satellite is newly launched, to establish the standard magnitude (half-illuminated at 1000 km distance). The brightness is assessed by comparison with a number of stars near the track. 'About half a magnitude brighter than star A, nearly the same as star B, just a little fainter than C . . . ' you say to yourself and then look up the magnitude of A, B and C in the Bečvář star atlas, or better still the *Smithsonian Star Catalog*. You should achieve an accuracy of 0.5 quite easily, and experts claim to achieve 0.2.

Even if there are some variations in brightness, it is always useful to record the average magnitude of any newly launched object. The international designations, allocated rapidly on the basis of the first radar sightings, are quite often wrong: for example, there are problems when one rocket launches a large number of satellites, or when two objects are rather similar, such as the Molniya launchers and launcher rockets left in low orbit after sending off the Molniya into its highly eccentric orbit. In all these situations, visual observations are crucial in deciding the identity of objects.

More interesting than the steady satellites are those that vary regularly or irregularly in brightness. If the variation is irregular, you have to try to record it as best you can, but it is a difficult task. Probably the most efficient procedure is to speak a running commentary on the magnitudes into a tape recorder. Afterwards you can plot the magnitudes against time and see whether there is any regularity underlying the apparent chaos. Some observers like making such detailed analyses, but others find it frustrating when there seems to be no 'rhyme or reason' for the variations.

Satellites that vary regularly in brightness are more satisfactory targets, because you can record the flash period (the interval between one maximum of brightness and the next), and also the magnitude at maximum and minimum brightness. At its faintest the satellite may be invisible, but that information too can be useful. In estimating the flash period you should, if possible, time a number of cycles. With a split-action stopwatch you may time, say, 10 and 20 cycles and compare the values, thus checking your accuracy. Satellites with flash periods between 1 and 5 seconds can be timed most accurately. If the flash period is less than $\frac{1}{2}$ sec, you may miscount the number of flashes. If the flash period exceeds 10 seconds, you can usually time only a few cycles and you may not be able to judge the exact moment of the maximum or minimum brightness. If the maximum of brightness is a real 'flash' and can be accurately timed, you can achieve remarkable accuracy in estimating the flash period. For example, timing 50 cycles at 56.7 seconds, and 100 cycles at 113.2 seconds, gives a flash period of 1.132 seconds, with an error of 0.002 sec if the timings were accurate to 0.1 second.

Most spinning satellites tend to lose their rotational energy as time passes, as a result of eddy currents arising from the satellite's passage through the Earth's magnetic field, and also (if the satellite is low) from aerodynamic forces at perigee. Both these phenomena can be studied by analysing a satellite's spin rate, and it is fascinating to time the flash period for several nights in succession and see for yourself the gradual increase in the period.

Fluctuations in brightness are usually caused by the satellite tumbling end-over-end in orbit, and the flash period, if it is regular, is normally about half the time for one complete tumble. If the flash period is long, however, it may differ appreciably from the tumbling period, because the maximum brightness recurs before the satellite completes one revolution. Fig 44 gives an example of a rocket which turns through only 140°, rather than 180°, between two successive flashes. If the satellite's height is y km and its flash period is P seconds, the fractional difference between flash period and tumbling period can be as much as $0.6P/y$: so if $P = 20$ seconds and $y = 200$ km, the difference is 0.06, ie 6%. However if the flash period is short, say 2 seconds, and the height greater, say 1000 km, the difference is negligible (0.1%). The tumbling period can then be accurately estimated from visual observations.

If a number of observers, geographically separated, record the maximum and minimum magnitudes of a rotating cylindrical satellite on the same transit, the direction of its axis of rotation in space can be determined. If the observations are kept up for several months, the movement of the spin axis in space can be followed, and compared with that expected on theoretical grounds. So, from simple magnitude estimates, interesting research projects can develop.

Fig 44 *A rotating cylindrical satellite usually appears brightest when the rays of the Sun (here shown at a depression angle of 30°) are reflected from its surface directly to the observer, as at points A and C. A complete flash period occurs between A and C, but the satellite only rotates through 140°, not 180°, so that the tumbling period is 18/14 times greater than the flash period.*

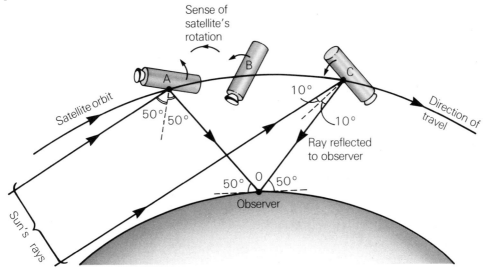

The brightest flashers of all are satellites sporting flat panels — usually covered with solar cells for generating power. These flat surfaces can produce powerful flashes, and some Molniya satellites can be seen with the naked eye at a distance of 40000 km in flashes which occur about once a minute. A flash of magnitude 3 at 40000 km would be seen at 1000 km (though over a much smaller area) as of magnitude-5. So the flashes of Molniyas at perigee, and other satellites with flat panels, such as Seasat, can outshine Venus at its brightest.

Why observe?

From time to time visual observers are taunted by remarks from ill-wishers: 'What's the use of visual observing? Surely it can be done much more accurately with cameras and lasers?' When I wrote the earlier version of this book in 1966, this second question had some validity, because the Baker-Nunn cameras of the Smithsonian Astrophysical Observatory were in full operation: they tracked about 20 satellites, and visual observations of these satellites were merely a 'gilding of the lily', though observations of other satellites were needed. Now, to keep up the clichés, 'the boot is on the other foot': the Baker-Nunn cameras no longer operate, except in support of laser tracking, which is itself confined to about a dozen special satellites fitted with the necessary corner-reflectors. There are now only a few large cameras operating, and visual observations are essential to provide the necessary numbers and geographical spread of observations if satisfactory orbits are to be determined from the camera observations. The second of the two questions is therefore not only ill-natured but also ill-conceived: the first question alone deserves an answer.

The analysis of changes in the orbits of chosen satellites has proved to be a powerful and inexpensive method for studying two quite different aspects of our earthly environment — (1) the gravity field and shape of the Earth, and (2) the behaviour of the upper atmosphere, particularly its density and winds. The procedure is to choose promising satellites; to drum up as many observations as possible; to determine the orbit from the observations as accurately as possible; and then to examine the changes in the orbits, as explained in Chapters 9 and 10. The observations come from many sources, including military radar (if available), camera observations (if any) and visual; of these, the visual observations are the most numerous and geographically varied, and their average accuracy is almost as good as the radar observations. The orbit analysts are always crying out for more visual observations: never yet have they had too many, and, because of the lack of observations, orbits are usually determined only once a week instead of daily as they might be. Being an orbit analyst myself, my hope in writing this book is to attract some (accurate!) new observers to swell the totals of visual observations, improve the orbits, and assist the geophysical researches.

Orbit analysis is in itself enough to justify organized visual observing, but there are many other uses, which I shall rapidly run through.

The most immediate use is in correcting the predictions. An observer can, within minutes if necessary, telephone the prediction centre with a correction for a badly predicted satellite, and this can be added to the ansaphone corrections, thus ensuring that most observers and the Hewitt cameras will see it on the next clear night. This not only produces more observations but also raises the morale of observers who may have been discouraged at failing to see the satellite. Morale is very important in visual observing: it is always easier to watch television than go out in the cold to observe.

A second useful service for skilled observers is to look for difficult new satellites, to decide whether they are bright enough to be worth including on the list for orbit analysis. This is exacting and valuable work — exacting because you may have to search many times, and valuable because the inclusion of an invisible satellite on the observing list may seriously damage observers' morale.

The third use of visual observations is in identifying newly launched satellites, and estimating their size and general shape. As mentioned on page 102, a skilled visual observer can often identify the type of satellite and rocket by looking at them, rather like plane-spotting but more skilful because there is no outline visible. Sometimes the international designations of a satellite and its rocket are interchanged, and a visual observer can easily spot this mistake if the satellite is stabilized in attitude and steady in brightness, while the rocket is spinning and fluctuates.

The next use is similar but more creative — keeping track of objects that evade the US Air Force radars and are not catalogued in the official lists. In recent years, the prime examples have been fragments from the 30-m diameter Pageos balloon, which are large orbiting pieces of plastic, about twice or perhaps five times the size of a plastic raincoat. These are often visible to the naked eye, but do not show on radar screens. Visual observers, led by Pierre Neirinck at the Appleton Laboratory, Slough, have kept their own orbits for these objects — an alternative culture, as it were, outside the official listings, which include only objects detectable by radar. Some of the objects unseen by radar can be quite substantial: in April 1965 two large fibreglass shields separated from the Snapshot satellite (1965-27A) and were visible to the naked eye in flashes but went unrecorded by radar for some time. Since then, there seems to have been a 'gentleman's agreement' among the launching organizations not to launch objects invisible to radar.

Visual observers can also see satellites in close formation that are otherwise recorded as a single object, being unresolved by radar. Two satellites 100 metres apart at a distance of 500 km subtend an angle of $0.01°$ and can be seen as a pair with 11×80 binoculars or any larger instrument. Though separated satellites usually drift apart (and are then given separate international designations), some are designed to fly in formation: thus each of the US Navy's Ocean Surveillance satellites has a trio of 'SSU' satellites flying in formation near it — an example being 1976-38C, D and J. (The SSU satellites are several kilometres apart, but separations of 100-500 metres may be needed in other applications.)

Visual observers can also determine the satellite's spin rate and spin axis direction, from observations of the flash period, and the maximum and minimum brightness, as described in the previous section.

The records of flash period are also often valuable to orbit analysts. If a cylindrical rocket keeps spinning at a rate of at least one revolution per minute, its effective cross-sectional area is likely to remain near its average value. But if the rocket ceases to rotate, the cross-sectional area could change greatly, and misleading results could arise if the rocket were used for long-term studies of air density. So, before embarking on their work, scientists studying air density take a searching look at the records of spin rate produced by visual observers.

The flash period has also been called 'the golden key to spin-ups': it is the best guide to unexpected increases in spin rate. Most satellites gradually tire of tumbling, as magnetic or aerodynamic forces whittle away their rotational energy. But some satellites or, more often, rockets surprise us with a sudden 'spin-up', usually caused by a more-or-less explosive release of compressed gas or unburnt propellant. This is also very important in orbit analysis, because the release of gas may alter the orbit and invalidate the analysis. The orbit of Cosmos 54 rocket, 1965-11D, for example, suffered a 5 km change in perigee height between January and June 1966 when the flash period decreased from 19 seconds to 4 seconds.

The last use of visual observations that I shall mention (though I know there are others) is the most spectacular — observing satellites during their final fiery plunge through the lower atmosphere, when they shine brightly by their own light, whether in eclipse or not. A large decaying satellite (see Fig 12, page 21) puts on a good firework display for about 2 minutes, during which time it travels about 600 km, and descends from a height of 90 km to about 30 km. The firework display can be seen easily over an area about 1000 km long and 600 km wide, which is $\frac{1}{800}$ of the Earth's surface. A satellite coming down in daylight, or on a completely cloudy night, will probably be unseen; so an observer who is out watching at the right time only has a chance of perhaps 1 in 2000 of seeing any particular satellite decay.

Tracking during re-entry will therefore only be possible if numerous observers are available, and that means volunteer visual observers. Also the human eye is the best instrument for observing re-entry at night, because the track is not accurately predictable and wide-angle vision is needed initially to sense the glow.

Nearly all the fragments recovered have been found as a result of visual observation of the decay. The Moonwatch organization at the Smithsonian Astrophysical Observatory was particularly successful in this work and arranged 're-entry patrols' for many decaying satellites until its much-regretted closure in 1975. The fragments of Sputnik 4 (Fig 13, page 22) and many other satellites were found after Moonwatch observers saw the satellite burn up. Inevitably, however, the majority of decays are seen only by non-observers, and it is notoriously difficult to establish the track from their conflicting reports. Decays are usually too spectacular to be conducive to accurate observation, but if you do see one, try to note and identify any bright star near the track: 'it passed about 5° above (or below) the Pole Star' would be a very useful fix. There are two decays of large satellites every week, so there is always a chance that you may see one. Nearly every week someone somewhere in the world sees a satellite decay. So keep your eyes open!

Observing feats

The end-product of visual observing is a list of times, right ascensions and declinations. These are stored on computer cards or magnetic tape; and serve as the raw data for the science of orbit analysis. It is all very clinical and impersonal.

But the observations are made by real people, devoted observers who forgo easy entertainment (like watching television) to engage in strenuous efforts to catch a satellite on the wing and pin-point its position. The observer may have to endure bitter cold, lose sleep, or wrestle with clouds; and the slightest inefficiency or inattention leads to complete failure. And all this work is quite unpaid. Every productive observer deserves a medal for public service. I cannot award medals but I should like to mention some outstanding observers, with apologies to those who are unfairly ignored. It is difficult to make a selection because complete and up-to-date worldwide statistics on the numbers and accuracy of observations are not available.

In the USSR one of the leading stations in the 1960s was at Enisseisk, where a team of three usually tried to observe 50 or 60 transits per night. The leader of the group, Valentin Vorotnikov, personally observed 40 transits of 26 different satellites on 28 March 1965. Another very active observing centre was at Lvov University in the Ukraine, where Alexander Logvinenko had made observations on over 200 satellites by 1966.

In the USA visual observing was organized by the Moonwatch Division of the Smithsonian Astrophysical Observatory, Cambridge, Massachusetts. Since Moonwatch was closed down in 1975, observing in the USA has returned to cottage-industry status, relying on keen local groups. One of the leading Moonwatch observers was Dr Arthur Leonard of Sacramento, California, who succeeded in recovering lost satellites on many occasions. In July 1961 his team saw 54 fragments from the rocket of Transit 4A soon after it exploded in orbit. Dr Leonard continues to observe and has made more than 10000 observations.

The Head of Moonwatch in the mid 1960s was William Hirst, and after his retirement in 1968, he began a new observing career in South Africa, with a telescope of aperture 13 cm, magnification 20, and a field of view of $2\frac{1}{4}°$. Using the telescope as a theodolite, he has made more than 13000 observations with an accuracy of about 0.03°.

You might think that the cloudy skies of Britain would discourage observers. On the contrary: the frequent clouds allow observers to maintain their enthusiasm without being grossly overworked. The world's leading observer is Russell Eberst of the Royal Observatory Edinburgh, who began observing in 1958 and had made more than 90000 observations by 1982. He observes from his home in Edinburgh with 11 × 80 binoculars and achieves an average accuracy of about 0.03°. His record night was 15-16 September 1966 when he made 127 observations on 40 satellite transits, and his highest monthly total, 1145 observations, was made in November 1977. Not far behind Eberst in numbers is David Hopkins of Bournemouth, who generally uses 10 x 70 binoculars and made 65000 observations between 1968 and 1982, including 1324 during August 1976, when the sky was nearly cloudless in southern England. Among other skilled and prolific British observers may be mentioned David Brierley of Malvern, Peter Wakelin of Sunningdale and Michael Waterman of Camberley, who has specialized in identifying new launches. The most accurate of the British binocular observers is Stanley Milbourn of Copthorne, Sussex, who uses mounted 11 × 80 binoculars and achieves an accuracy between 0.01° and 0.02°, thanks to the careful use of very faint stars. Britain's leading telescopic observer is Gordon Taylor of the Royal Greenwich Observatory, Herstmonceux. He made 9000 observations on about 750 satellites between 1957 and 1982, and in recent years he has been observing from Cowbeech, Sussex with a telescope of aperture 50 cm, achieving an accuracy between 0.01° and 0.02°.

Great skill, enthusiasm and pertinacity are needed to achieve results like those described in the previous paragraphs. Though a small (and unpaid) 'profession', satellite observing has, in my opinion, higher standards than most other professions.

7
With Camera and Laser

Star of the Earth and diamond of the night.

Erasmus Darwin, *The Botanic Garden* (1792)

I have so far concentrated on visual observing of the 'stars of the Earth', because anyone who is keen can become a visual observer and make a real contribution to scientific research. A much more accurate method of optical tracking is available, however: photography. The snag is that it is also much more expensive in equipment and labour. Even more accurate than photography, and currently the ultimate at optical wavelengths, is laser tracking, when a laser beam illuminates a satellite fitted with corner reflectors — a genuine 'diamond of the night' — and the satellite reflects back the light, so that the time of travel out-and-back can be accurately measured. In this chapter I take a rather superficial look at camera and laser tracking, and describe some existing cameras and new designs, but with no attempt at the kind of detail lavished on visual observing.

Photography

Taking photographs of satellites as they travel past the stars is an obvious refinement of visual observing and should also be much more accurate. Instead of merely estimating the fractional distance between two stars, you can measure it up on a photographic plate; and an electromechanical timing system can replace the fallible human hand and eye. But cameras are subject to the same limitations as visual observers over eclipse and prediction: indeed a camera is more vulnerable to prediction errors, because most cameras are set pointing towards the predicted track and a satellite running early or late may not cross the field of view at all.

Photographic tracking is simple enough in principle. The aim is to record on a photographic plate (or film) both the image of the satellite and the images of the stars among which it passes. The exact positions of the stars are known from star catalogues, so the exact direction of the satellite as seen by the camera can be determined by carefully measuring the distance of the satellite from several of the reference stars on the photographic plate.

Exact timing is also essential. A satellite may have an angular rate of travel of $1°$ per second; so if the observation is accurate to $0.001°$ in direction, as is possible with a camera, timing needs to be accurate to 0.001 second, or 1 millisecond. Otherwise the timing error will be dominant and the laborious work of measuring the plate will be wasted. Usually the timing is performed by suddenly interrupting the tracking, for perhaps 2 seconds or perhaps $\frac{1}{2}$ second (depending on the satellite's angular speed), by closing a fast shutter, or using a rotating chopper, as in the Hewitt camera. The intervention of the shutter or chopper is registered simultaneously by the timing unit, usually a quartz crystal clock checked continually against radio time-signals.

There are several possible techniques for photographic tracking, and in illustrating them I shall assume that three breaks are made in the track at regular intervals, perhaps every 2 seconds.

The first possible way of operating is with the camera stationary. The stars then trail slowly across the field of view, at not faster than $\frac{1}{4}°$ per minute, while the satellite moves rapidly across at perhaps $20°$ per minute, and has its trail broken on the three occasions when the shutter closes. The image on the photographic plate looks rather like Fig 45(a). The short broken lines represent the tracks of the stars, here the seven stars of the Plough; the satellite's track is the long broken line. The time at the beginning of each break is known; so the satellite's position and the star positions at the beginning of each break can be measured on the plate.

A second mode of operating is to make the camera follow the stars, just as an astronomical telescope does. The stars then appear as dots and the satellite track as a broken line. This is the situation shown in Fig 45(b). Again the satellite's position relative to the stars at each of the three breaks can be measured.

These methods seem fine, but both suffer from one possible snag: unless the satellite is bright, its track may be extremely faint or quite invisible. The slow-moving stars have a much better chance of impressing their image on the photographic plate than the fast-moving satellite.

The only ways out of the difficulty are expensive: you must either use a much larger camera, or devise a mechanism to allow the camera to follow the satellite. To rotate the camera at the correct speed about the correct axis is a formidable task. The camera has to be cradled in an elaborate mounting, with a three-axis or even a four-axis gimbal system and mechanical drives to turn the camera at the correct rate, which must be calculated beforehand. If all these requirements are satisfied, the result is as shown in Fig 46(a). The satellite appears as a solitary point-image among

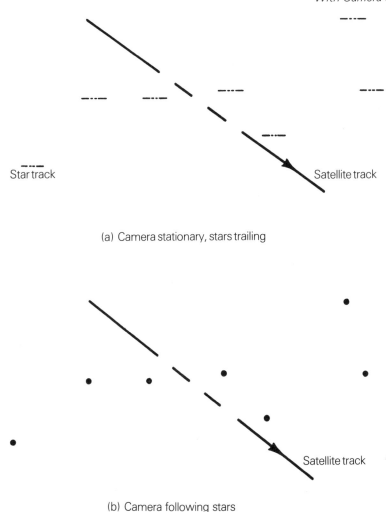

(a) Camera stationary, stars trailing

(b) Camera following stars

Fig 45 *Two possible methods of photographic tracking. See also Fig 46.*

rapidly moving stars. Because the stars move rapidly, faint ones are not recorded at all; but this does not matter if there are enough bright stars near the satellite's track.

Making the camera follow the satellite is not entirely satisfactory, however, partly because faint stars are not recorded, and partly because you sometimes need the eye of faith to be sure that the point image *is* the satellite and not a spot on the film.

To secure the best of both worlds, the oscillating mode of tracking can be used, in which the camera first follows the stars and then follows the satellite. Fig 46(b) shows the situation when the stars and satellite are bright, so that their trails can be

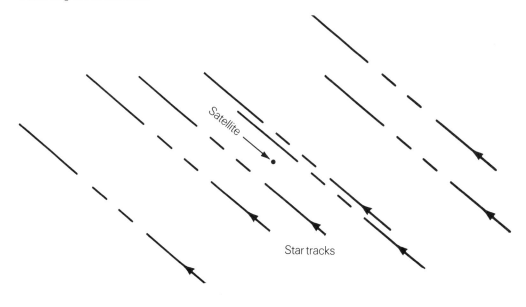

(a) Camera following satellite

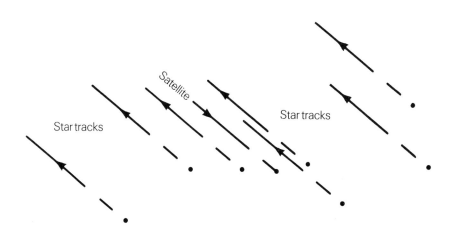

(b) Camera following first stars, then satellite

Fig 46 *Two further methods of photographic tracking.*

seen. Suppose however that the satellite and all the stars in Fig 46(b) are faint, so that their trails are invisible, but the point images show up. Then the point images accurately represent the position of the satellite and stars at the moment when the method of tracking changes: you secure one observation in conditions when the other methods of operation would yield no observations at all.

These four basic techniques can be altered in many ways. For example, the camera can be rocked from side to side at approximately the same rate as the satellite is moving. A satellite image is recorded only when the camera is swinging in the same direction as the satellite moves.

All these photographic methods are upset if the satellite fluctuates so greatly in brightness that it is invisible to the camera over a large proportion of its track. For if the break in the trail occurs when the trail is not visible, the break cannot be located; and if the break comes while the satellite is visible, can you be sure it is caused by the shutter rather than by the satellite becoming invisible? Because of this limitation, cameras are best used on satellites that are fairly constant in brightness.

Large cameras

Trying to record images of faint satellites with small cameras is more of a sporting pursuit than a routine operation, and in practice the major contributions have been made by large cameras, of aperture greater than 20 cm (8 inches, for the benefit of olde-time astronomers!). It is just because these cameras are so expensive to manufacture and to operate that visual observing has retained its vitality. If small cameras could be easily used, the visual observers could each be supplied with one.

I shall describe four large cameras: the Baker-Nunn camera of the Smithsonian Astrophysical Observatory, Cambridge, Massachusetts, which was the world leader in photographic tracking for many years; the British Hewitt camera, the most accurate in the world; the Russian AFU-75, which in recent years has made more observations than any other; and the new US Air Force instruments which began operating in 1982, though it is arguable whether they should strictly be called 'cameras' at all.

The Baker-Nunn camera, shown in Fig 47, was designed in the mid 1950s by James G. Baker and Joseph Nunn, specifically for satellite tracking, and has been most successful in that role. The aperture, ie the diameter of the cylinder of light utilized by the camera, is 51 cm (20 inches), but much of the bulk and weight (nearly 3 tons) come from the three-axis gimbal system, which allows the camera to follow almost any likely satellite track in the sky. The camera can be swung round manually while the satellite is observed through a side telescope, but it is usually moved automatically.

The camera is of a modified Super-Schmidt type, and its optical design is shown in Fig 48. The light enters through the correcting system of lenses on the left, and is focused by the pyrex mirror (on the right) on to the film, which is fed through the

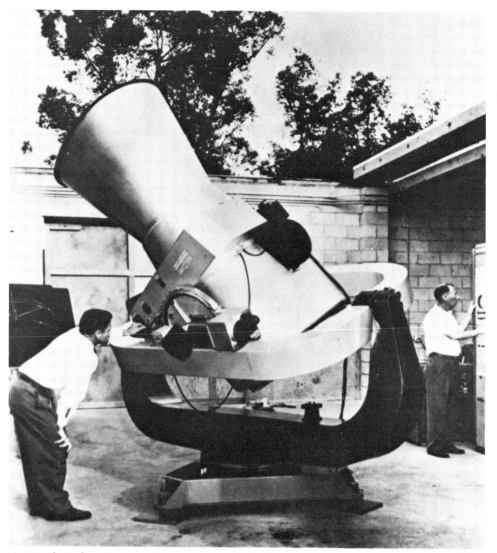

Fig 47 *The Baker-Nunn tracking camera of the Smithsonian Astrophysical Observatory. The camera has an aperture of 51 cm and can photograph satellites down to magnitude 12 with a directional accuracy of 3 seconds of arc. See also Fig 48.*

centre of the camera, in the focal plane. The mirror is 76 cm (30 inches) in diameter; the aperture and focal length are both 51 cm, so that the system is *f*1. The focal surface is approximately spherical and the film is held under tension across it.

The field of view of the camera is 30° along the satellite's track and 5° perpendicular to it. The scale of the image is such that stars 1° apart in the sky are about

1 centimetre apart on the film. With the best film-reading machines, the positions of the satellite and stars can be read correct to about 2 microns (0.002 mm) corresponding to an accuracy of 0.0002°. However, because the tracking of the satellite is not perfect, and because the film does not stretch quite evenly, the accuracy attained in practice for satellites is usually about 0.001° (3 seconds of arc).

A complete exposure usually lasts for less than 4 seconds and a rotating shutter near the focal plane makes breaks in the trails of the stars (or satellite) five times at precisely determined moments, so that a Baker-Nunn exposure usually has five breaks in it. At one of the breaks, a flash-light is activated and records the time registered on a clock.

The timing system for the Baker-Nunn camera is based on a crystal clock, synchronized with standard time signals. The timing accuracy for satellite observations

Fig 48 *Cross-section of the Baker-Nunn camera. The light enters on the left, passes through the lenses of the correcting system to the primary mirror and is reflected back on to the film at the focal surface.*

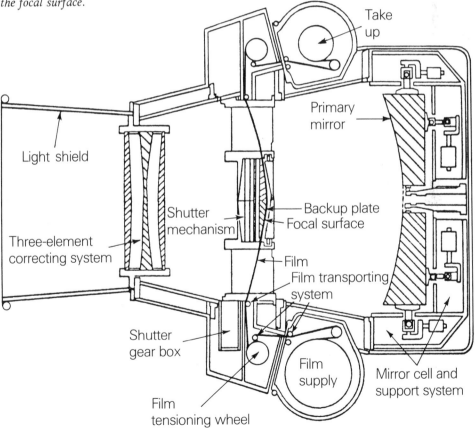

is usually about 2 milliseconds, that is, 0.002 second. This is well matched to the positional accuracy of about 0.001°, since the fairly high satellites which best suit the Baker-Nunn camera rarely move faster then $\frac{1}{2}$° per second and cover less than 0.001° in 0.002 second.

The Baker-Nunn cameras need accurate predictions and thrive on faint satellites in low-drag orbits. With a 20-second exposure, the camera can photograph stars down to magnitude 14, and satellites of magnitude 12 are well within reach. Attempts to track high-drag satellites have not been so successful, merely because of the expense and effort of frequently updating predictions for a worldwide network.

Though large and costly, the Baker-Nunn camera has triumphantly justified itself: the cameras have made more than a million observations of high accuracy. The Smithsonian Astrophysical Observatory began operating the cameras in 1958 and throughout the 1960s maintained a network of twelve cameras at various sites distributed round the world. During the 1970s the use of the cameras decreased, not through any fault of theirs but because the funds were diverted elsewhere. The US upper-atmosphere researchers, being physicists rather than mathematicians, decided to rely on expensive instruments in spacecraft to measure upper-atmosphere properties 'on the spot', rather than analysing orbits obtained from observations. The US geodetic researchers, striving for higher precision, were not satisfied with the 10-metre positional accuracy offered by the Baker-Nunn, and moved into lasers with a potential accuracy of 10 cm or better. So now the Baker-Nunn cameras are used only in support of the laser programme.

The British Hewitt camera (Fig 49) was designed in the late 1950s by Joseph Hewitt, working at the Royal Radar Establishment, Malvern. Two cameras were manufactured, by Grubb Parsons Ltd. The camera was intended to record the track of the Blue Streak ballistic missile, and when the Blue Streak programme was cancelled in 1960, this beautiful instrument was left without a role. The Hewitt camera was well suited to satellite observation, so one camera was installed at a site near Evesham, where it began operating in 1963. In 1967 the two cameras were purchased by the Ordnance Survey for use in the European satellite triangulation by observation of the Pageos balloon; the second camera was set up at the Earlyburn outstation of the Royal Observatory Edinburgh. For the next ten years the cameras gave good service, in the European satellite triangulation and in observing satellites for orbit analysis in geophysical researches. But by 1975 the Ordnance Survey's role had changed, with emphasis on selling maps rather than on geodetic science, and their need for the cameras had vanished. So in 1978 the Ordnance Survey kindly presented the cameras to the University of Aston in Birmingham, where the Earth Satellite Research Unit of Dr C. J. Brookes had won a grant from the Science Research Council for optical tracking and orbit analysis of satellites. The intention was to move the Edinburgh camera to the southern hemisphere, and in 1980 this move was made: the camera was set up at an excellent site at Siding Spring, New South Wales,

Fig 49 *The Hewitt camera developed at the Royal Radar Establishment, Malvern, and now owned by the University of Aston, Birmingham. The camera has an aperture of 61 cm and can photograph satellites with a directional accuracy of 1 second of arc. The photograph shows the camera at Siding Spring in Australia, which began operating in 1982. See also Figs 50 and 51.*

Australia, where observations began in 1982. The other camera was due to be moved in autumn 1982 to an excellent new site at the Royal Greenwich Observatory, Herstmonceux. In 1982 both cameras were operating as well as ever: they were still the world's most accurate satellite cameras, and they have recorded more than 4000 satellite transits, with 5 to 10 observations per transit.

The Hewitt camera is similar to the Baker-Nunn in several respects: but it is not a tracking camera, so that the mounting is simpler. The Hewitt camera is a field-flattened Super-Schmidt, $f1$ and of 61-cm (24-inch) aperture. Its field of view is 10° in diameter. Although its aperture is appreciably greater than the Baker-Nunn's, the camera cannot record such faint satellites because it remains stationary during each exposure, instead of tracking the satellite. A satellite of magnitude 10 can be recorded if it is 4000 km distant and moving at 0.1° per second; but if the satellite moves 10 times faster, at 1° per second, it has to be 10 times brighter, about magnitude $7\frac{1}{2}$, if it is to be recorded.

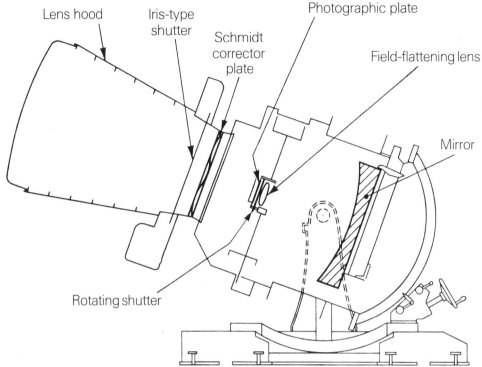

Fig 50 *Cross-section of the Hewitt camera. During a satellite observation the iris shutter is open and light passes through the corrector plate to the mirror, where it is reflected on to the field-flattening lens and then to the flat photographic plate (or film fixed to a glass plate).*

The layout of the Hewitt camera is shown in Fig 50. The light enters through the Schmidt corrector plate and is focused by the 86-cm diameter mirror on to the photographic plate. A lens in front of the plate ensures that the rays converge on to a plane rather than a curved surface as in the Baker-Nunn camera. The Hewitt camera is often operated in 10-second bursts: the main shutters, of iris type, are opened for 10 seconds while a rotating chopper, rather like the blade of a car-engine fan, produces breaks in the trail at intervals of 1 second. But it is a feature of the camera that dozens of different modes of operation are possible: however irregular the satellite's behaviour, the camera probably has an answer to it.

The directional accuracy of the Hewitt camera is 0.0003° (1 second of arc), which is three times as good as the Baker-Nunn. This is chiefly because the Baker-Nunn camera records the images on film which may be slightly distorted because it is stretched over the curved focal surface, whereas the Hewitt camera was designed to record the images on a flat glass photographic plate — though since 1977 these expensive plates have been replaced by film fixed on to clear glass plates, with no loss in accuracy. Fig 51 shows an example of a Hewitt camera photograph. The timing system is similar in principle to that of the Baker-Nunn, and in 1980 a new control and timing system for the cameras was manufactured to replace the original electronics, of 1960 vintage.

The third camera on my list is the Russian AFU-75 (Fig 52) designed in 1965 at the tracking station of Riga University by K. K. Lapushka and M. K. Abele. The camera has a four-axis mounting and stands on a specially designed equatorial platform. Like the Baker-Nunn camera, the AFU-75 can be rotated to follow the satellite. The optical system is a seven-lens arrangement of aperture 21 cm and focal length 74 cm. The

Fig 51 *A photograph taken by the Hewitt camera at Evesham. The long trail is the track of the satellite 1962 alpha upsilon as it passed through the constellations of Taurus and Auriga on the evening of 17 January 1963. The numerous short lines are star trails during the time when the shutter was open and the satellite was crossing the field of view. Each of the star trails has a break, which is scarcely discernible, at the time of the long gap in the satellite's trail. The three dots on the left and two on the right of each star trail result from short exposures of the stars, made before and after the satellite observation, to determine the direction in which the camera is pointing.*

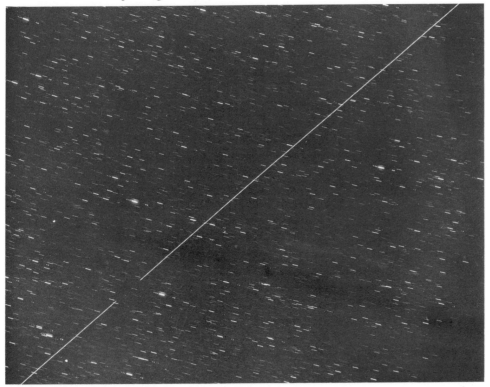

Fig 52 *The AFU-75 camera of the USSR. The camera has an aperture of 21 cm and can photograph satellites down to magnitude 10 with an accuracy of 3 seconds of arc.*

field of view is $10° \times 14°$ and the images are recorded on film 19 cm wide. There is also a guiding telescope of aperture 12 cm for visual control of the tracking if necessary. Timing is by crystal clock, oscillograph and radio receiver. Satellites down to magnitude 10 can be recorded. A full description of the AFU-75 is given in the article by A. G. Massevitch and A. M. Losinsky listed in 'Further Reading'.

The AFU-75 cameras began operating about 1970 at the following sites: Riga, Uzhgorod, Zvenigorod and Pulkovo in the USSR; Ondrejov (Czechoslovakia); Sofia (Bulgaria); Baja (Hungary); Ulan-Bator (Mongolia); Havana (Cuba); Cairo (UAR); Afghoi (Somalia); the Kerguelen Islands (Indian Ocean); and Mirny, in the Antarctic. The locations of the cameras have been changed from time to time, but large numbers of observations continue to be made. Independent assessments of accuracy suggest a figure of about $0.001°$, though it is possible that the instrumental accuracy is better than this.

There are many other large cameras in use round the world, but it is not possible to describe them all here: for details, see 'Further Reading'. These other cameras include the Zeiss SBG camera (GDR) of aperture 42 cm; the Antares (France) of aperture 30 cm; a Schmidt-type camera in Finland of aperture 34 cm; and the largest USSR tracking camera, the VAU, of aperture 50 cm. Also, any large astronomical telescope can be adapted for satellite photography, though very accurate prediction is needed because of the small field of view of such telescopes.

The last camera on my list could be called either 'a camera for the 1980s' or 'not a camera at all'. I refer to the 'ground-based electro-optical deep space surveillance system' (GEODSS), being developed by the United States Air Force, for operational use during the 1980s. There are plans for five GEODSS stations, possible locations being White Sands (New Mexico); Maui (Hawaii); Taegu (South Korea); the eastern Atlantic region; and the Indian Ocean. It is planned that each station will have three telescopes (or 'cameras'), two of 1.02 m aperture (Fig 53) with a 2° field of view, and one of 38 cm aperture with a 6° field of view. The expected accuracy is about 0.0005° (2 seconds of arc).

Fig 53 *A telescope of the US Air Force's ground-based electro-optical deep space surveillance system (GEODSS). The aperture is 1.02 m and the 4-axis mounting weighs 9 tons. The object is to detect all satellites at heights between 5000 and 36000 km. The system started operating in 1982.*

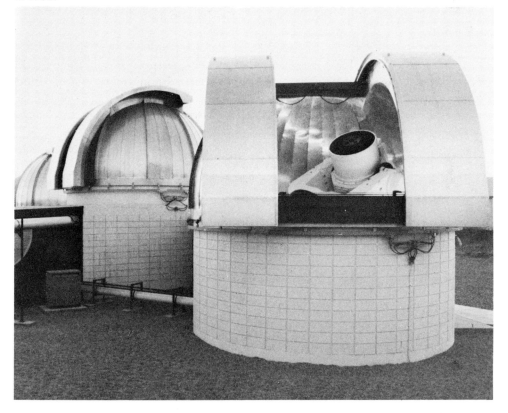

The GEODSS system does not use photographic film: instead, sensitive vidicon tubes and radiometers, with computer control, will produce images in real time, which can be monitored on visual display units. If these displays are photographed to provide a permanent record, the instrument becomes virtually a camera; but otherwise it is perhaps better called a satellite tracking telescope with instantaneous visual display — the ultimate in visual observing, with camera facilities added.

The GEODSS telescopes have 4-axis mountings weighing 9 tons and are intended for space surveillance at heights of 5000 km to 36000 km. The aim is to be able to search the whole sky in about one hour and to detect any satellite more than 30 cm in diameter. Since most high satellites remain above the horizon for an hour or more on a near-overhead transit, they should be detected on any transit when the sky is clear and dark. Although optical tracking is 50 times more difficult at 36000 km range than at 5000 km, radar tracking is 2500 times more difficult: that is why optical techniques have now won favour with the USAF.

Small cameras and kinetheodolites

There is no essential difference between 'large' and 'small' cameras, though the latter usually have lenses rather than mirrors and of course cannot record such faint satellites.

The faintest satellite within the grasp of a fixed camera depends on (a) the aperture, (b) the focal ratio f (the focal length divided by the aperture), and (c) the angular speed of the satellite. With fast film, a camera of aperture 10 cm, and $f1$, would record a fast satellite ($1°$ per second) of magnitude 3.5 or brighter. A camera of similar aperture and $f2.5$ would record fast satellites of magnitude 2.5 or brighter, or slow satellites ($0.1°$ per second) of magnitude 5 or brighter.

One small camera that has been widely used is the Russian NAFA 3c/25, with aperture 10 cm, focal length 25 cm and a very wide field of view, $30° \times 50°$. The accuracy is about $0.001°$ in direction and 3 milliseconds in time, which rivals the larger cameras.

Another much-used fixed camera is the US Minitrack Optical Tracking System (MOTS) installed at NASA Minitrack radio tracking stations. The aperture is 20 cm, the focal length 1 m, and the field of view $11° \times 14°$. The capability is about the same as the NAFA 3c/25.

Many other useful cameras have been developed in various countries, often with ingenious methods for improving performance, eg by rocking or rotating the camera or plate; but it is not appropriate to go into further detail here.

The NAFA and MOTS cameras are not so good as the naked eye in detecting fast satellites, though better than the naked eye for very slow ones. Not many bright satellites have been launched in recent years, so small fixed cameras are of limited value.

All the cameras so far mentioned 'hitch their wagons to a star', because the position of the satellite's image is measured relative to nearby star images on the film. A camera does not need solid earth, and could be on a balloon in mid ocean, if you carefully determined the position of the balloon.

But there is one photographic instrument, the kinetheodolite, that keeps its feet firmly planted on Earth and could operate even if all the stars were wiped off the face of the sky. In a kinetheodolite the satellite's direction is determined from the angle between its image and the cross-wires defining the axis of a small telescope; and the direction of the telescope itself is measured relative to the Earth, in terms of its elevation above the horizon and its azimuth.

A kinetheodolite (Fig 54) consists of a telescope cradled in an accurate and rather heavy mounting, which allows the telescope to be pointed in any direction. It can be rotated horizontally, so as to look towards any point of the compass, and it can be elevated to any angle above the horizon. Dials are provided which show accurately the compass point (azimuth) and the elevation angle. Two observers control the direction in which the telescope points, each looking through a smaller side-telescope

Fig 54 *A kinetheodolite in action. This shows the instrument used at the Royal Observatory Edinburgh in the late 1960s: the observers are B. McInnes (left) and R. Eberst (right). A similar instrument at the South African Astronomical Observatory made observations on 15000 satellite transits between 1969 and 1981.*

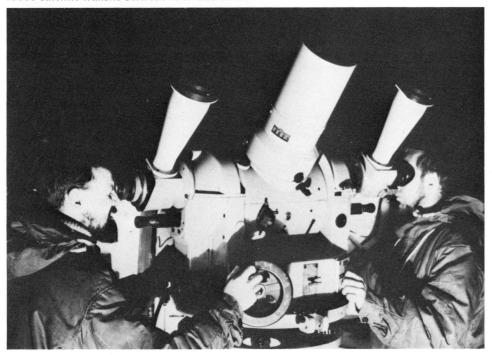

parallel to the main telescope: one observer moves the kinetheodolite in azimuth, ie decides the compass direction; and the other observer moves it in elevation, deciding what angle of elevation is needed. Kinetheodolites are designed primarily for tracking missiles on test ranges, so that skilled observers find no difficulty in following satellites, which are comparatively slow moving. The task of each observer is to keep the satellite as close as possible to the centre of the field of view, which is defined by the cross-wires. While the observers track the satellite in this way, many photographs, of perhaps 0.2-second exposure-time, are being taken inside the kinetheodolite. These record (1) flashlight photographs of the azimuth and elevation dial scales, (2) the time, which is fed in electronically, and (3) the image of the satellite, showing its position relative to the cross-wires. If the tracking were perfect, there would be no need for (3). When the film is developed, the exact time at the beginning of each exposure can be read off, and the azimuth and elevation can be suitably corrected to allow for the displacement of the image from the centre of the cross-wires. A hundred or more observations can be made on a single transit of a satellite; normally there is no need for more than five or ten observations per transit, but the capability of taking 100 observations is valuable when satellites are in their last orbits or re-entering the atmosphere.

Observations with a kinetheodolite can attain an accuracy of about 0.005° in direction and 0.01 second in time. The disadvantage of a kinetheodolite is that it can only photograph bright satellites, of magnitude 4 or brighter, because it is a 'small camera' of aperture about 10 cm. However, telescopes of larger aperture can be fitted, to improve performance, and there is also scope for 'photovisual' operation, with the satellite being observed visually, and the azimuth and elevation recorded photographically as usual, whenever the satellite is seen to be at the intersection of the cross-wires.

Kinetheodolite observations were successfully made throughout the 1960s and 1970s. One instrument in particular has an outstanding record. This is an Askania kinetheodolite which, after being 'retired' from work on missile ranges of the Royal Aircraft Establishment, began a new career of satellite tracking at the Royal Greenwich Observatory in 1965, and was moved to the South African Astronomical Observatory in 1969. There, thanks to the devoted attention of Walter Grimwood, the rather ancient instrument was kept in full operation for twelve years and recorded observations on 15000 satellite transits before being closed down in 1981.

Photoelectric observation

In the previous chapter I described how visual estimates of satellite brightness could be used to analyse the rotational motion of a satellite. Again, however, you can do much better than the human eye with expensive equipment, in the form of a tracking camera and photoelectric cells. Very rapid variations in brightness can then be accurately recorded: there is a wealth of detail, and the analysis of such a trace is quite difficult.

Photoelectric measurements offer rich possibilities for future research: it is a field through which Dr Kenneth Kissell has cut a solitary furrow with a skilful hand. But the field still remains largely unploughed, and so here I can do no more than point to its promise.

Lasers

The mention of lasers, the trendy gadgets of the seventies, makes some people glow with pleasure and others wince at the dangers to eyesight. Normal light arises from a disorderly vibration of atoms, and the resulting light waves are all mixed up. The essential feature of a laser is that it emits light waves which are coherent — all alike in phase and wavelength. Such well-disciplined waves can be dragooned far more effectively than a mere rabble, and can be projected as an exceedingly narrow beam, rather like a searchlight beam but much narrower.

If a satellite is to be tracked by laser, the laser beam must be directed very accurately, to ensure that it illuminates the satellite; so the prediction of the satellite's track must be very precise. The satellite also has to play its part in the operation: it must be specially designed to send back the light along exactly the line on which it arrived. To achieve such perfect reflection, the surface of the satellite has to be studded with corner-cube reflectors — made by slicing a corner from a transparent

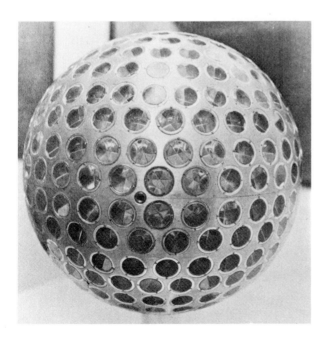

Fig 55 *The laser geodynamic satellite Lageos, which was launched in 1976 into a circular orbit at a height of 6000 km inclined at 110° to the equator. The satellite is 60 cm in diameter, with a mass of 411 kg and its surface is studded with 426 corner reflectors.*

125

cube, so that any ray of light entering the sliced face will be turned through exactly 180° after two internal reflections. (The cat's-eye studs on roads work on the same principle.) The laser geodynamic satellite Lageos shown in Fig 55 is a good example of such a 'diamond-studded' reflector. If the laser light is sent out in pulses, the time taken for the light pulse to go out and come back can be measured, and the satellite's distance can be determined since the speed of light is accurately known. Thanks to sophisticated timing devices and pulse-chopping techniques, the distance can be determined with an accuracy of a few centimetres.

The first satellite to be fitted with laser retroreflectors was Beacon Explorer B, 1964-64A, followed by BE-C, 1965-32A, which is still in use. Between 1965 and 1981, fifteen further satellites with reflectors were launched. So there were enough

Fig 56 The satellite laser tracker at the Royal Greenwich Observatory, Herstmonceux, instal-led in 1982 and due to begin operating in 1983. The receiving telescope is of 50 cm aperture. The system is designed to have an accuracy in distance measurement of 2 cm. (A smaller dome would have been adequate, but it was convenient to use an existing empty dome.)

satellites available for the first laser tracking operations, which began in the late 1960s, with ruby lasers. At first the accuracy of distance determination was about 1 metre, but this figure was gradually reduced during the 1970s, by technical improvements in shortening and chopping the pulses, by changing to Neodymium-Yag lasers, and by advances in timing accuracy. A number of lasers now operating have accuracies of 10 cm, and the new lasers of the 1980s should achieve an accuracy of about 2 cm.

The USA was the first country to develop satellite laser trackers, being followed by France and the USSR. During the 1970s the US stations set up by the Goddard Space Flight Center and the Smithsonian Astrophysical Observatory continued to play a leading role, but laser trackers were at work in many other countries too. Britain was late, but in 1982 a satellite laser tracker was installed at the Royal Greenwich Observatory, Herstmonceux, in cooperation with the University of Hull, in a project financed by the Science and Engineering Research Council. The installation is shown in Fig 56.

This laser is of advanced design and will serve as an excellent example of a laser of the 1980s. The laser is a mode-locked Neodymium-Yag system operating on a wavelength of 530 nanometres (the second harmonic), which is in the green region of the spectrum. The pulse length is about 150 picoseconds (a picosecond being a million millionth of a second, or 10^{-12}s). There are 10 pulses per second, each with an energy of 30 millijoules — which would raise the temperature of 1 gram of water by only 0.007°C. The laser pulses are transmitted through a refracting optical system with an aperture of 10 cm, and there is a reflecting telescope of aperture 50 cm for receiving the returning photons. The detection and timing system is capable of responding to single photons, and the time of travel to the satellite and back can be measured accurate to about 100 picoseconds. Since light travels 300000 km in 1 second, an error of 100 picoseconds in timing corresponds to an error of 3 cm in distance. The British laser tracker has a radar slaved to it which will detect any aircraft about to cross the laser beam, and the beam will be switched off automatically if this happens, to avoid the (very slight) danger that aircraft passengers might look at the laser.

Because of its superb accuracy, laser tracking is of high importance for the future. But it has two important limitations. First, the return signals cannot be detected at all unless the satellite is fitted with corner reflectors. In 1982 only ten of the 5000 satellites in orbit had usable corner reflectors, and laser tracking is confined to these favoured few satellites. The second problem is that the position of the satellite must be very accurately predicted; otherwise the narrow laser beam will miss it altogether. For a satellite like Lageos at a height of 6000 km, the divergence angle of the beam has to be less than 0.01°, to obtain adequate return signals, and predictions of a similar accuracy are required. For satellites free of air drag and well observed, such accuracy in prediction is just possible; but satellites with perigee lower than 500 km have not so far been regularly tracked by laser, because air drag prevents accurate prediction. But if the sky is dark and the satellite is not in eclipse, it should be feasible to track such satellites visually and adjust the aiming of the laser accordingly.

Like all other forms of optical tracking, a laser needs clear skies; but laser tracking is not limited by eclipse or daylight. At night a satellite in eclipse can be tracked just as easily as a sunlit one (unless visual tracking is needed); and satellites can be detected by laser in daylight, though not so easily as at night.

The 2 cm accuracy of laser tracking should greatly improve knowledge in many areas in geophysics, including the variations in the Earth's rotation and pole position, the drift of the continents, the shape of the geoid, the possible contraction or expansion of the Earth, and so on. This new knowledge should be of social benefit in improving methods of earthquake predictions. So, if the welfare of humanity and the advance of science are regarded with favour, laser tracking should have a bright future.

8
Electronic Eyes

Greet the unseen with a cheer!

Robert Browning, *Asolando* (1889)

In the last four chapters we could see what we were about. Now it is time to throw away our eyes and greet the unseen radio waves. With optical tracking, by eye or by camera, you have to wait, sometimes for weeks or months, until the satellite allows itself to be illuminated by the Sun against a dark sky. Radio waves free you from this dependence on solar illumination and a dark sky, and from worries about clouds. These are the great advantages of using radio waves in satellite tracking.

The radio waves can be utilized in various ways, and I shall first divide them into two groups, *radio* and *radar*. In *radio* tracking, the satellite transmits radio signals, which are received and interpreted by instruments usually (but not necessarily) on the ground. In *radar* tracking, on the other hand, the satellite is passive and is illuminated by a powerful transmitter on the ground. The echoes returning from the satellite are then received and interpreted at ground receiving stations. (Laser tracking is obviously rather like radar in some ways.)

Radio tracking offers three main possibilities: measuring the direction of the signals arriving from the spacecraft (interferometry); measuring their frequency (Doppler); and measuring their time of travel (ranging).

Radar tracking needs a different subdivision, into 'normal' radar, with the transmitter and receiver on the same site; and 'radar interferometry', when the receivers are interferometers well separated from the transmitter.

I shall discuss electronic tracking under the five headings specified in the last two paragraphs. My survey is inevitably rather superficial, covering general ideas rather than going into technical detail.

Radio interferometers

'Interferometer' is a very long word for a simple device: it measures interference between radio waves received at two aerials. Basically an interferometer consists of two dipole aerials — two iron rods will do — horizontal and parallel to each other, placed at the same height above the ground and at an accurately known distance apart, 4 wavelengths in the example shown in Fig 57.

The interferometer operates by comparing the satellite's radio signals, as received at the two aerials: whenever the signals reinforce each other, ie are 'in phase', the distance from the satellite to aerial A must either be the same as the distance from the satellite to aerial B, or, more likely, differs from it by an exact number of wavelengths, as shown in Fig 57. Since the satellite is at least 100 km away, the radio waves arriving at the aerials can be regarded as a parallel beam.

In Fig 57 the diagram on the left applies when the aerial A is 2 wavelengths further away from the satellite than B, the line BC being perpendicular to AC. Since the length AC is half the length AB, the triangle ABC is half an equilateral triangle and the angle BAC, the angle of elevation of the satellite above the horizon, is 60°. Similarly the diagram on the right of Fig 57 applies when the difference in distance is 1 wavelength instead of 2. The satellite is then at an elevation of about 75° — the exact figure is 75.52°. If the satellite is at the same distance from both aerials, its elevation is 90°. As the satellite passes from overhead away into the distance, an interferometer can therefore record the times at which AC is 0, 1, 2, ... wavelengths, giving elevations of 90°, 75.52°, 60°, ...

As its name implies, an interferometer also records the occasions when the waves

Fig 57 *Principle of the radio interferometer. The two aerials A and B are sited 4 wavelengths apart. When the phase difference between the waves arriving at the two aerials is equivalent to 2 wavelengths (left), the satellite is at an elevation of 60°. When it is one wavelength (right), the elevation is 75°.*

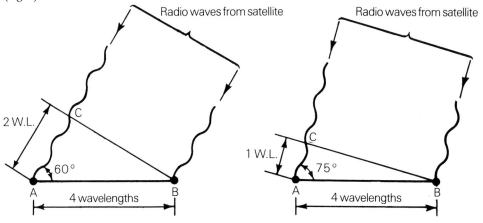

completely interfere with each other, ie when the distance AC is $\frac{1}{2}$, $1\frac{1}{2}$, $2\frac{1}{2}$, ... wavelengths. These times of zero signal can be found more accurately than the times of maximum signal, and provide a series of readings at elevations of 82.8°, 68.0°, 51.3°, ... if the aerials A and B are 4 wavelengths apart, as in Fig 57. To secure readings at closer intervals, the aerials can be spaced more widely: if they are 10 wavelengths apart, zero signals occur when the satellite is at elevations of 87.1°, 81.4°, 75.5°, ...

So far we have been assuming that the satellite is in the same vertical plane as the aerials. This would happen if the aerials were in a north-south line and the satellite chanced to be exactly north or south. But such good fortune is rare, and a second pair of aerials is usually needed, in an east-west line. With 'crossed' aerials, one pair north-south and the other west-east, an interferometer can completely define the direction of a satellite which is transmitting radio signals of a suitable wavelength. Any ambiguity — is one aerial $\frac{1}{2}$, $1\frac{1}{2}$, $2\frac{1}{2}$ or $22\frac{1}{2}$ wavelengths further from the satellite than the other? — can be resolved by using a second pair of aerials with a different spacing. Interferometers can provide a series of observations as the satellite passes over.

Radio interferometers have been of great service in tracking the satellites launched by the US National Aeronautics and Space Administration. For twenty years a system of twelve 'Minitrack' interferometer stations, distributed round the world, has faithfully recorded the transits of all NASA satellites on a frequency of 136 megahertz (MHz): Fig 58 shows the Minitrack station in England, at Winkfield near Ascot. But the Minitrack stations are being 'phased out' in favour of tracking from orbit by the Tracking and Data Relay Satellite System scheduled to be launched by the Space Shuttle in 1983.

Radio waves bend slightly as they pass through the ionized layers of the upper atmosphere, and this limits the accuracy of interferometers. The bending is minimized by using a high frequency and making observations at high elevation (above 60°). High-elevation Minitrack observations have a directional error of about 0.02°, similar to the best visual observations.

Radio Doppler tracking

If you live in a part of the world where express trains advertise themselves with a whistle rather than a klaxon, you can scarcely fail to notice how the pitch of the whistling note falls as the train passes by. The whistle is high-pitched as the train approaches and low-pitched as it recedes. This phenomenon is known as the Doppler effect, and the change in pitch is exactly proportional to the speed of the train, so that if the pitch is measured, the speed of the train can be found. That is not all: for the nearer you are to the track, the quicker the pitch will change; so you can also find your distance from the railway line.

In Doppler tracking, the radio signals from satellites can be regarded as nothing

Fig 58 *The Minitrack radio interferometer station at Winkfield, near Ascot, England, where NASA satellites on a frequency of 136 MHz were tracked daily from 1960 to 1979.*

more than a very high-pitched whistle. Imagine a fast train travelling along a straight viaduct, whistling, and imagine yourself in the valley below, some distance from the track, armed with a tape recorder. You record the high-pitched whistle of the approaching train and the lower-pitched whistle as it recedes. Arrange for an accomplice to do the same, further away from the track. When you meet and play over the two recordings, you will know you were nearer the train, because the pitch changes more quickly on your recording. And if you measured the pitch exactly, you could tell how far away you each were.

The situation is the same for satellites, except that they move in ellipses rather than along straight lines, and the great virtue of the Doppler method derives from the fact that the detailed variation in the frequency of the radio signals received at a ground station depends on the size, shape and orientation of the orbit. So, if you measure the changes in frequency accurately enough, you can determine the satellite's orbit. The total change in frequency is less than 10 kHz for a satellite transmitting at 100 MHz, so the satellite transmitter must be very stable in frequency to avoid introducing spurious variations into the received signals.

Radio Doppler tracking has been exploited with great success in the US Navy's

Navigation Satellites, often called 'Transit'. The system comprises five satellites in circular polar orbits at heights near 1100 km, each broadcasting very stable signals on the linked frequencies of 150 and 400 MHz. The orbits of the satellites are determined (from Doppler observations) with an accuracy of about 1 metre, and, since air drag is slight at a height of 1100 km, it is possible to predict the orbits for 24 hours ahead with high accuracy. Each satellite can therefore be supplied with information about its own orbit and can continually broadcast its exact position in space, and its orbital parameters, to anyone on the ground with a suitable receiver. Knowing the satellite's position and orbit, the detailed variation in the frequency of the signal received at the ground station can be analysed to determine the position of the ground station.

Thanks to the advances in computers and micro-miniaturization in recent years, the equipment for receiving the signals from the Transit satellites has now been reduced to a portable 'black box', which is ideal for survey purposes. The survey team takes the 'black box' to a particular site, receives data from Transit satellites on several passes, giving the position of the site accurate to about 1 metre at best; and then the team moves on to the next site. For more details, see *Satellite Doppler Tracking*, listed in 'Further Reading'.

An accuracy of 1 metre is better than most navigators require, and many users of the Transit satellites opt for a simpler and less accurate receiver. In 1981 there were about 10000 such users, including ships at sea and oil-rigs as well as surveyors on land.

Like the interferometer, Doppler tracking is troubled by the distortion of radio waves in the ionosphere, and that is why the Transit system uses the two frequencies 150 and 400 MHz. Generally, the errors are inversely proportional to the frequency: so they can be assessed and removed by intercomparison.

Listening in

When satellites are equipped with radio transmitters, it is usually because someone wishes to send back information, perhaps giving the results of environmental measurements made by the satellite, or perhaps just reporting the state of health of the instruments aboard. So the satellite usually sends back a coded stream of information (telemetry) as well as emitting a continuous signal for use in tracking. The reception and interpretation of the coded telemetry signals at their 'home' station is essentially a private transaction between the satellite and its owners, and is outside the scope of this book. But if the coded signals are strong enough, they can be received and interpreted by other stations; and such 'eavesdropping' qualifies as a method of observing, especially when the signals are used to register and identify a new launch.

This kind of detective work, particularly with Cosmos satellites, has become the speciality of 'the Kettering Group', a band of amateur radio enthusiasts round the world, whose observations are coordinated by Geoffrey Perry, senior physics teacher at Kettering Boys' School in England. Perry began his monitoring of transmissions from the Russian reconnaissance satellites and manned spacecraft in the early 1960s. They transmitted powerfully on command at frequencies near 20 MHz, and he gradually learnt how to interpret their signals, being able to detect and identify new launches within minutes and determine recovery times accurately.

As the recoverable Cosmos and Russian manned satellites are bright, and easily seen by visual observers, quick detection of their launch can provide predictions for the visual observers later the same day: a prompt reaction is needed because these satellites remain in orbit for only two weeks at most. The recoverable Cosmos satellites are also important because they are so numerous — there were about 500 of them up to 1982. At the end of their missions they advertise their return to Earth by switching on a recovery beacon, which usually broadcasts the letters TF, TK or TL: the actual recovery time, when the signals fade, is about 6 minutes later. Since the USSR has not revealed the telemetry procedures, this creative interpretation of what might otherwise have been written off as gibberish has won wide acclaim for Perry and his group.

Simple receivers suffice to pick up the powerful signals from the Russian satellites, which are usually at heights near 200 km. But it is a different story with the communication satellites at heights of 36000 km: large aerials, like the big dishes at Goonhilly in Cornwall, are needed to record reliably the messages or television pictures from the communication satellites.

Even more difficult is the tracking of space vehicles flying off to other planets. Again large dishes are needed, and the 76-metre radio telescope at Jodrell Bank in England was often used for this purpose in the 1960s. Since these space vehicles are not Earth satellites they will receive no further mention here.

However, there are quite a large number of transmitting satellites in lower orbits that *can* be detected by amateur radio observers. Finding and identifying these transmitting satellites is rather fascinating, and also useful, because many are not to be found in the official lists. Some of these 'rogues' are satellites that have been told to switch off, but disobey orders; others are satellites that have 'returned to life' after a long silence, on being heated during a lengthy spell in sunlight undisturbed by eclipse. Gregory Roberts of the South African Astronomical Observatory has specialized in tracing such satellites, and in 1981 there were 57 transmitting satellites on his list. To these must be added communication satellites and those transmitting only on command, which might bring the total up to about 100, that is about 2% of the 5000 objects in orbit.

Radio ranging

As the accuracy of clocks improved in the 1960s, the question arose, 'Would it be possible to measure accurately the travel time of a radio signal from satellite to ground?' If so, the distance of the satellite from the ground station could be determined. The idea appealed to Dr Roger Easton, of the US Naval Research Laboratory, designer of the Minitrack system, and he saw the possibility of fixing position on the Earth by measuring your distance from two or more such satellites. Easton successfully pursued the idea, which was adopted by the US Navy and Air Force in the form of the Navstar satellites (*Nav*igation *s*ystem using *t*ime *a*nd *r*ange). There were six Navstar satellites (see Fig 59) in orbit by 1982 and the system is due to become fully operational in 1987, with 18 satellites in circular orbits of period 12 hours (height 20000 km) inclined at 63° to the equator.

The idea of Navstar is simple. The satellite carries an accurate clock and has a code to identify the time at which it sends out a signal. The receiving navigator, on a ship, say, has a synchronized clock and notes the time of reception. The difference gives the time of travel, and therefore the satellite's distance from the navigator. The satellite's orbit, and hence its position when it issues the signal, is accurately known. The navigator then takes a reading on another Navstar satellite — there should always be four above the horizon when the sytem is complete. This gives the distance of the ship from a second known point. The two measurements define the geographical position of the ship. Some computation is required, of course, but it is all standard and is preprogrammed into microchip computers within the 'black box' containing the receiver and clock, which is light enough (15 kg) to be carried by a (strong) individual. By using three satellites, you can find your position in three dimensions, and by using a fourth satellite you can eliminate the clock error. Repetition of the measurements will give your velocity accurate to 0.1 metre per second, which is particularly useful for aircraft. The frequencies are 1575 MHz and 1227 MHz. With simpler equipment, using only the first frequency, accuracies of about 10 m are expected. With more complex equipment, accuracies better than 1 m should be achieved.

Fig 59 *A Navstar satellite. There are plans for 18 of these in 12-hour orbits in the late 1980s. If you have a suitable receiver, you will be able to fix your position on Earth to 10 m with their aid, or 1 m with the most accurate equipment.*

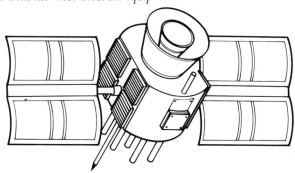

Signals from Navstar received at known stations can also be used to determine very accurate orbits. Used like this, Navstar is similar to laser tracking, except that only half the out-and-back travel time is measured.

Navstar shows the possibilities of 'pure' radio ranging: but why not combine it with Doppler, by placing a transmitter or transponder in the satellite and measuring both the time delay and the frequency shift in the received signal? This was done in the Goddard Range and Range Rate system (GRARR) of the NASA Goddard Space Flight Center, developed in the 1960s. In determining the satellite's orbit it is obviously an advantage to have both types of measurement available, and so it turned out: GRARR was most successful.

With GRARR, the signals were received at ground stations. But low satellites are only visible from ground stations for short spells (5–10 minutes) on each orbit, so that numerous ground stations are needed. You can do better by using a synchronous satellite instead of a ground station. The synchronous satellite is in effect 'tied' to the rotating Earth: it is a tracking station stuck at the top of a pole 36000 km high, with the advantage that it can see each low satellite for 50% of its orbit instead of only 5–10%. The method has been tried successfully with the ATS 6 satellite as the tracking station, and will operate widely in the Tracking and Data Relay Satellite System (known as TDRSS, pronounced 'tea-dress'). Two of these satellites, stationed at longitudes 41°W and 171°W, will be used to track all NASA spacecraft after 1983.

A similar system can be used to detect small changes in the orbital path due to anomalies in the gravitational field of the Earth. The velocity of one satellite relative to another can now be measured correct to a few millimetres per second, and this can give the detailed acceleration history of a satellite. For example, the Soyuz-Apollo was tracked from ATS 6 in this way, and unusual accelerations were detected as the spacecraft passed over the Himalayas.

Other radio techniques are possible, such as measuring the velocity difference between two satellites in identical low orbits a few hundred kilometres apart. Gravity anomalies will be revealed by their differing accelerations. But we are now straying into specialized areas where very few satellites are to be found, and we must return into the mainstream of observation.

Normal radar

The simplest radar tracking system consists of the familiar movable saucer-shaped dish, which sends out radio waves in a near-parallel beam and detects the small fraction of them reflected from any distant object — aircraft, satellite or perhaps a bird. If such an echo is detected, the direction of the reflecting object is the direction in which the dish is pointing, and its distance is known from the time taken by the signal to travel out and back (at the speed of light, 300000 km per second).

A radar observation therefore has the advantage that it gives both an angular direction and a range measurement; and radar can operate when optical methods fail because of clouds and daylight, and radio methods fail because the satellite does not transmit. Indeed radar has so many advantages that you may well ask, 'why hasn't radar superseded all other methods?'

There are five reasons, which I once called 'the five plagues of radar', though partial cures for some have now been found. The first and worst problem is cost. At a distance of 1000 km, a visual observer with 7 × 50 binoculars, costing £30 or so, can detect and track satellites down to 1 m in diameter. To do the same with radar calls for a dish about 15 m in diameter costing perhaps half a million pounds, with a professional team to run it and heavy demands for electrical power. Because of the cost, it is no surprise to find that nearly all existing radars are operated by military organizations.

The second problem with radar is that the performance deteriorates rapidly as distance increases. If the distance doubles, the intensity of the radio waves falling on the satellite is reduced to a quarter, and the proportion of the reflected waves collected by the dish is also reduced to a quarter. So when the distance is doubled, the satellite needs to have an echoing area 16 times greater if it is to be detected. A radar that can detect spherical satellites down to 1 metre in diameter at 1000 km range would not be able to detect anything of diameter smaller than 25 m at 5000 km range; this is an unimpressive performance, because a spherical satellite 25 metres in diameter would be of visual magnitude 5 at 5000 km, and just about visible to the naked eye. The poor performance of radar at great distances has impelled the USAF to go over to optical tracking for deep space surveillance (see previous chapter).

The third snag is that radar cannot detect satellites made of material like fibreglass or plastic. Several fragments from the plastic mylar Pageos balloon, often bright enough to be visible to the naked eye, have been circulating in orbit for years undetected by radar.

The fourth limitation is the directional accuracy: since a large radar distorts a little as it rotates, it cannot compete with a camera in directional accuracy. Very few radars are more accurate directionally than a good visual observer, and it is a general rule to take the same directional accuracy for visual and radar observations.

The accuracy of radar can be improved by using a narrower beam, but this leads to the fifth problem — the need for accurate predictions. If the beam is $\frac{1}{4}°$ wide, the predictions must also be accurate to $\frac{1}{4}°$, whereas a camera or visual observer can tolerate errors up to 2° or even 4°, depending on the field of view. This problem can be alleviated by switching from wide to narrow beams.

If I have 'piled on the agony' in detailing these five problems, it is because radar is (apart from these snags) the ideal method of tracking, being able to function in daylight, in cloudy weather and in the absence of any cooperation from the satellite.

By far the most important system of normal radar tracking operates as part of the Space Detection and Tracking System (Spadats) of the North American Air Defense Command (NORAD). The complete NORAD detection system has three main components: there are optical sensors, at present Baker-Nunn cameras, to be replaced soon by GEODSS; there is the US Navy's Navspasur system, to be described in the next section; and there are several powerful radars in various parts of the world. The most recent published map of these stations (dating from 1977) is shown in Fig 60; but there have been some changes since then.

Three important radars in the NORAD system are the Ballistic Missile Early Warning stations. The first is at Clear in Alaska; the second is at Thule in Greenland; and the third is at Fylingdales Moor, Yorkshire, England, where the spherical domes protecting the large dishes have become a familiar feature of the landscape (see Fig 61). Though the primary role of these stations is to detect ballistic missiles as they hurtle through space, their radars inevitably also detect any satellites that are large enough and near enough. That includes the majority of all the objects in orbit, and the observations are immediately transmitted to NORAD.

The radars at Eglin, Florida and Shemya, Alaska, are of a different type: they are huge phased-array installations, looking rather like the face of a cliff, but made up of numerous small 'antennae' which are 'steered' electronically, so that there are no mechanical moving parts to limit the speed of scan. These radars are designed to detect satellites down to 10 cm in diameter at distances out to 4000 km. There are also large phased arrays, 32 m × 32 m, known as Pave Paws, at Beale Air Force Base,

Fig 60 *The radar and optical sensors of NORAD's Space Detection and Tracking System, as existing in 1977. There have subsequently been some changes.*

Fig 61 *General view of the Ballistic Missile Early Warning System station at Fylingdales Moor, Yorkshire, England, operated by the Royal Air Force. The spherical domes protect the radar dishes from wind, rain and snow.*

California and Otis AFB, Massachusetts. In addition, as the map shows, there are smaller but still powerful radars at Antigua, Ascension Island, and elsewhere.

About 20000 satellite observations per day flow from these and other sensors to the NORAD headquarters inside Cheyenne Mountain near Colorado Springs. The headquarters building there is a three-storey 'office-block' with a staff of nearly a thousand, and the building fits closely inside a huge cave carved out of Cheyenne Mountain: there are 400 metres of solid granite above the cave. To reach the NORAD headquarters, you have to drive along a tunnel about 500 metres in length and through enormous steel doors that can be quickly closed. The buildings, made mainly of steel (with no glass), are mounted on huge spiral mechanical springs about 3 metres in diameter. The Cheyenne Mountain Complex is designed to remain operational after an attack by nuclear weapons: the air is filtered, communications are diversified, and a month's supply of food and other essentials is carried. The Cheyenne Mountain Complex is among the wonders of the modern world and may well stand for millions of years as an eloquent monument to the bygone human passion for war, to be visited in future more peaceful days by human survivors of a nuclear holocaust or our successors of a different species.

Radar interferometer

Since pure radar tracking has limitations, why not combine the virtues of radar and the radio interferometer? The two essential features of such an instrument would be

(i) a very powerful transmitter of radio waves, and (ii) very sensitive detecting aerials, which pick up the reflected rays and indicate their direction. The transmitter and the receivers can, however, be physically quite separate instead of using the same dish; there may be several receiving stations instead of one; and the signals may be transmitted continuously, not in pulses, as is customary with normal radar.

The prime example of a radar interferometer system is the US Navy's 'Naval Space Surveillance System'. Navspasur, as it is usually called, was another of Dr Roger Easton's inventions. The idea was that a narrow fan-beam of radio waves, very wide in the east-west direction and very narrow in the north-south direction, should be broadcast by an extremely powerful transmitter in the southern USA. Any satellite crossing this vertical east-west 'fence' would reflect the radio waves, and the reflections could be detected at a number of interferometer stations on an east-west line across the continent. The system became operational in 1961 and has been working 24 hours a day ever since, feeding more than 10000 observations per day to NORAD in recent years.

The main transmitter is at Lake Kickapoo in Texas (with two subsidiaries at Gila River and Jordan Lake), as shown in the map, Fig 62. The main transmitter is probably the largest and most powerful in the world. It consists of an array of 2556 dipoles, extending in a straight line for more than 3 km, and radiates continuously with a power of 810 kilowatts on a frequency of 217 MHz (wavelength 1.4 m). The

Fig 62 *The disposition of the transmitters and receiving stations of the US Navy's Navspasur fence across the southern USA. Six interferometer stations detect reflections from satellites crossing the east-west fan-beam produced at Lake Kickapoo.*

Fig 63 *Aerial view of the Navspasur transmitter at Lake Kickapoo, Texas, looking south. The 2256 dipoles stretch for more than 3 km and the power radiated continuously is 810 kilowatts. The lower half of the photograph covers only the first two of the eighteen sections: the rest stretch away into the distance and a public road (just visible) crosses the transmitter half way along its length.*

141

Fig 64 *Ground-level photograph of the Navspasur transmitter at Lake Kickapoo, Texas.*

transmitter, shown in Figs 63 and 64, is in 18 sections, so that one section can be taken out for repair without much weakening the signals.

The radiation reflected from any satellite crossing the fence is received at the six interferometer stations marked in Fig 62, from San Diego in the west to Fort Steward in the east. All these stations are at latitudes near 33°N and they operate on the same principle as the radio interferometers already described: any satellite crossing the fence becomes an involuntary 'transmitter' of radio waves of 217 MHz frequency. A low satellite passing well to the east or west may be recorded by only the nearest one or two of the stations, but most satellites passing centrally are recorded by all six interferometers, and their nearly simultaneous observations can be combined to give a single accurate position in space for the satellite. The accuracy is about 200 metres for satellites at heights up to 500 km, with some increase in the figure as the height increases. The processing of more than 10000 observations per day is a formidable task, and is performed with the aid of three powerful computers at the Navspasur headquarters at Dahlgren in Virginia.

The great strength of Navspasur is its ability to measure the positions of all satellites crossing the fence (except for those too high to give a detectable signal). The Navspasur observations have proved to be of immense value in determining orbits, being not only reliable and accurate but also available at least twice daily on the satellite's north-going and south-going crossings of the fence. Again and again, the Navspasur observations have provided the 'solid backbone' for an orbit determination: without them the more intermittent optical observations would often have been virtually useless.

Optical and electronic tracking: a retrospect

In the past three chapters I have rattled through many methods of satellite observation with a rapidity that may have proved confusing. So it may be worth attempting the impossible task of summarizing them. Each method has its virtues and defects: that is why all have survived and no single technique has become dominant. The Table below is intended to give an approximate idea of their qualities.

Pros and cons of tracking methods

Method	Equipment and running costs	Satellites observable (per cent)	Accuracy at 1000 km (metres)	Weather needs
Visual, 11 × 80 binoculars	very low	75	200	dark, clear
Large camera (Hewitt)	medium	50	5	dark, clear
Large camera (Baker-Nunn)	medium	85	10	dark, clear
GEODSS telescopes	high	98	10	dark, clear
Laser tracker	medium	0.2	0.02	clear
Radio interferometer	medium	2	200	none
Radio Doppler (Transit) ⎫ Radio ranging (Navstar) ⎭	medium	0.3	1	none
Large radar (Shemya)	very high	90	200	none
Navspasur	high	90	200	none

A Table of this kind is provocative and will meet strong criticism from those knowledgeable in satellite tracking. Some items are certainly arguable, particularly the 'cost' column. The only cost no one can quarrel with is the 'very low' for volunteer visual observers, many of whom have made more than 10000 observations with 11 × 80 binoculars originally costing £100. Other costs are much more difficult to assess. For example, a radio interferometer is given as 'medium' cost and Navspasur as 'high', because it has a powerful transmitter plus six interferometers. But since Navspasur produces more than 10000 observations per day, its cost *per observation* is far less than that of a single radio interferometer or a camera. Also, you might call Transit cheap, and so it is for the user; but someone has to pay for launching the satellites. Such arguments can rage forever and a day without producing agreement.

9
Earth's Fair Form

Take any shape but that, and my firm nerves
Shall never tremble.

William Shakespeare, *Macbeth* (1606)

Who would have guessed that by sending satellites into space, observing them, and measuring their course across the sky, we should learn so much not about the secrets of space but about the shape of the Earth? The smart answer to this question is 'Christopher Marlowe', in *Tamburlaine the Great*, written twenty years before *Macbeth*. We 'can comprehend the wondrous Architecture of the world', Marlowe tells us, 'and measure every wandering planet's course'.

'Come my friends', wrote Tennyson, ' 'Tis not too late to seek a newer world'. He was right, and in this chapter I give one answer to the question 'Why observe?' by outlining the truly wondrous architecture of the newer world that has unfolded before our astonished eyes since 1960.

Before I start, I should make it clear that whenever I mention 'the shape of the Earth', I shall be referring to the average sea-level surface, carried through under the land in a logical fashion. This is the surface usually called the *geoid*. Mountains and other local irregularities do not enter the picture directly, though their presence may very slightly alter the geoid.

The great globe
Even our ancestors in Africa a million years ago could have noticed that an offshore island was visible from a cliff-top but not from the shore below. Some of the more imaginative may have had the idea of a spherical Earth, though most of them probably thought the Earth was basically flat.

144

But the idea of a spherical Earth is certainly of great antiquity, and was prevalent among the Greeks in the fifth century B.C. The Pythagoreans looked on the sphere as the perfect shape, an idea reinforced by the observation that the Sun and Moon appear circular. The Greeks may also have realized that the circular zone of darkness creeping across the Moon at an eclipse was the Earth's shadow; and they certainly heard tales of new stars, such as Canopus, that become visible on travelling south.

In the third century B.C., Eratosthenes of Alexandria used travellers' tales to measure the Earth's circumference, and quite accurately too. He noticed (or was told) that at noon in midsummer the Sun was directly overhead and would shine down a deep well, at Syene, or Aswan as it is now known; while at Alexandria, almost due north, the Sun was $\frac{1}{50}$ of a circle (7.2°) away from the vertical. So between Alexandria and Syene the direction of the Earth's surface changes by $\frac{1}{50}$ of a circle (if the Sun is assumed to be very distant), and the Earth's circumference is therefore 50 times the distance from Alexandria to Syene, a distance he estimated as 5000 stadia, because camels travelling on average 100 stadia per day took 50 days over the journey. So the Earth was 250000 stadia in circumference. The definition of the stadium is disputed, but under one definition his figure is within 1 % of the true circumference of just over 40000 km.

The idea of a spherical Earth persisted through the Middle Ages, being one of the few correct features of Aristotelian physics. Dante makes a spherical Earth and the Aristotelian spheres of Heaven the very basis of his poem the *Divina Commedia*. 'The great globe itself' was part of Shakespeare's imagery too, and he managed to describe the next advance in geodesy when he made Lear cry out 'Strike flat the thick rotundity o' the world!'

That is just what was needed, a flattening of the poles, as though giant hands pressing at the poles have slightly squashed the spherical Earth. The flattening f is defined numerically as the equatorial diameter minus the polar diameter (actually 42.77 km), divided by the equatorial diameter (12756.28 km). Thus the value of f is now known to be 1/298.25 or 0.003353.

Sir Isaac Newton was the first to make a numerical estimate of the flattening, in his *Principia* published in 1687. He used an ingenious argument, imagining one tube of water, or canal as he called it, running from the North Pole to the Earth's centre, and another from the centre to a point on the equator. Since water shows no wish to run from equator to pole (or vice versa) over the surface, the pressure exerted by each canal at the Earth's centre should be equal. The centrifugal effect of the Earth's rotation slightly weakens gravity in the equatorial canal, which therefore needs to be longer. After making the necessary calculations, Newton estimated the flattening as $\frac{1}{230}$. This is larger than the correct value of $\frac{1}{298}$, because he did not allow for any increase in density towards the centre. But for nearly a century his estimate remained the best and most soundly based.

Though no one could rival Newton as a theoretician, plenty were eager to confirm or refute his value by practical measurement. If the Earth was flattened at the poles, the distance corresponding to 1° of latitude would be greater near the poles than at the equator: in fact the distance is 111.69 km at the poles and 110.57 km at the equator. In 1736 the French Academy of Sciences sent out two expeditions to measure the length of a degree of latitude in Lapland and in Peru. After years of great hardships, followed by further years of dispute about the results, the estimates came out between 1/179 and 1/266, nicely bracketing Newton's value. During the next two hundred years the methods were continually improved and others were devised; and by 1957 the value generally accepted was 1/297.1, with a probable error of 0.4, as derived by Sir Harold Jeffreys and given in the third edition of his classic book *The Earth*.

Shaped like a pear

As we have seen in Chapter 4 (Fig 26), the gravitational pull of the extra bulge of material near the Earth's equator causes a satellite orbit to rotate about the Earth's axis, swinging to the west if, as is usual, the satellite is travelling eastwards. The rate of rotation is X degrees each day, where X is given in Fig 27, and 5 degrees per day is a typical value. Since the rotation rate is directly proportional to the flattening, a precise value of the flattening can be found by allowing the rotation to build up for 3 months or so, and accurately determining the rate of rotation. This method was used in 1958 with Sputnik 2 and indicated a flattening f near 1/298.2: this was appreciably different from the value of 1/297.1 previously favoured. According to the previous value of f, the equatorial diameter of the Earth was 42.94 km greater than the polar diameter. With the new value of f, 1/298.25, the figure is 42.77 km: the best-fitting spheroid has an equatorial diameter of 12756.28 km and a polar diameter of 12713.51 km.

So the satellite measurements in 1958 showed that the linear distance between the poles was 170 m greater than had been thought. This lengthens an air journey from equator to pole and back by $\frac{1}{2}$ km: but the airline operators had no need to raise their fares, because they had been flying the extra distance before without knowing it. But the discovery was a shock for geodesists, who had been working to an accuracy of 10 m and were upset to find their Earth in error by 170 m.

The new value for the Earth's flattening in 1958 was only the beginning of a rapid advance towards a full knowledge of the Earth's shape. The next step — while still retaining the simplifying assumption that the Earth is symmetrical about its axis — was to regard the profile of the Earth when sliced through the poles as made up of a circle modified by a whole series of more complex shapes, specified by what are called *zonal harmonics*. The *second* harmonic is a tendency towards an elliptic shape rather than a circle, as shown in Fig 65(a). This takes care of the Earth's flattening. The *third* harmonic corresponds to a tendency towards a triangular shape, as shown in Fig 65(b), or a pear shape as it is often called, because if the tendency were stronger

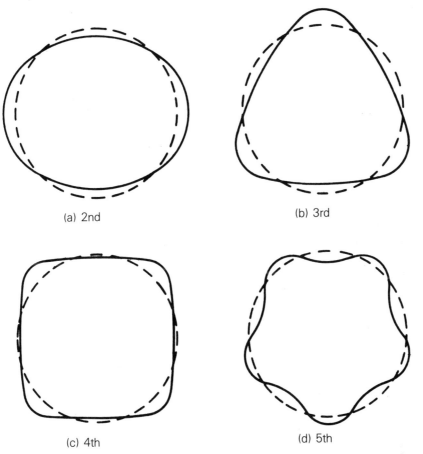

(a) 2nd

(b) 3rd

(c) 4th

(d) 5th

Fig 65 *Form of the 2nd to 5th zonal harmonics. The sections shown are slices through the poles and the shapes give variations with latitude averaged over all longitudes. The second harmonic corresponds to the Earth's flattening, the third expresses the 'pear-shape' effect, the fourth harmonic is square-shaped, the fifth has five 'petals', and so on.*

the shape would become concave and pear-like. The *fourth* harmonic, Fig 65(c), might be called square-shaped, as if to justify Donne's reference to 'the round Earth's imagined corners'; the *fifth* harmonic, Fig 65(d), has five 'petals'; and so on, in-definitely, although in practice the harmonics become very small for degree greater than about 20 or 30.

Each of these shapes gives rise to a certain pattern of gravity variation, which prints its own signature, as it were, on orbits. So, by analysing the orbits of a large number of satellites at different inclinations to the equator, we can determine the strength of each harmonic, if those beyond a certain degree are ignored. The even harmonics, those of degree 2, 4, 6, . . ., which are symmetrical about the equator, are evaluated from the observed rate of rotation of the orbital plane, as already explained.

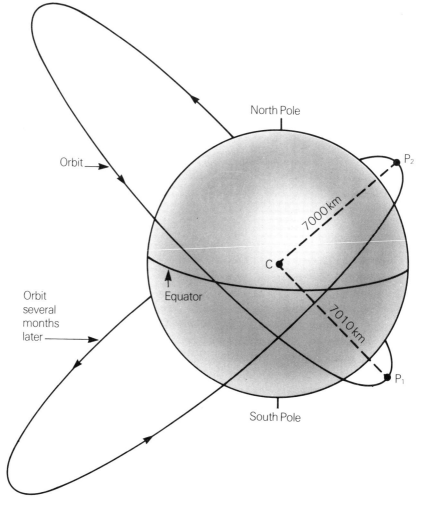

Fig 66 *As a result of the variations in gravity caused by the Earth's flattening, the perigee P of an orbit moves steadily, going from southern apex (P_1 in the diagram) to northern apex (P_2) and back again every few months. As a result of the asymmetry in gravity caused by the Earth being slightly pear-shaped, perigee is nearer to the Earth's centre when in the northern hemisphere. For an orbit with inclination 45°, as illustrated here, perigee is about 10 km nearer the Earth's centre when at P_2 than at P_1.*

But the odd harmonics, the 3rd, 5th, 7th, . . ., produce a quite different effect, namely a regular variation in the distance of perigee from the Earth's centre, as shown in Fig 66. The perigee P moves steadily from the southern apex of the orbit (P₁ in the

Fig 67 *The shape of the Earth averaged over all longitudes. The diagram is a slice through the poles and the solid line shows the height of the geoid, or sea-level surface, in metres, relative to a spheroid of flattening 1/298.25, shown as a broken line. The departure from the spheroid is greatly exaggerated, and the shape is of course not really concave at the south pole. (From D. G. King-Hele, C. J. Brookes and G. E. Cook, Geophys. J. Roy Ast. Soc., Vol.64, pp.3–30 (1981).)*

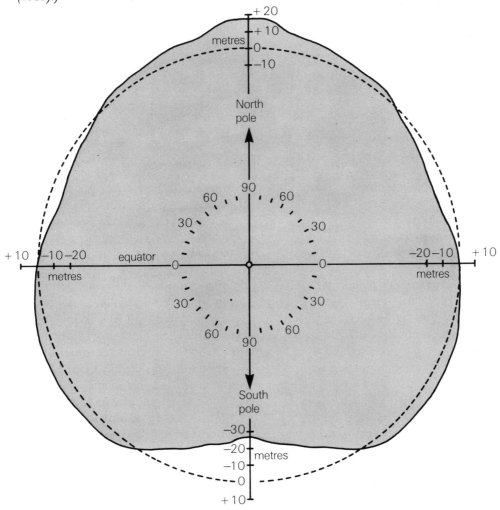

diagram) to the northern (P_2) and back again, every few months (the movement is caused by the gravitational pull of the even harmonics). As perigee moves, the gravitational pull of the odd harmonics alters the perigee distance CP. For an orbit of inclination 45°, the perigee is about 10 km nearer the Earth's centre when at P_2 than when at P_1. The 10 km expresses the total effect of odd zonal harmonics, and is different for other inclinations (7 km for 30°, and 19 km for 90°). By analysing numerous satellites at different inclinations, individual odd zonal harmonics can be evaluated, if those above a certain degree are ignored.

Putting together the values for even and odd zonal harmonics, we can find the detailed shape of the Earth sliced through the poles. This is shown in Fig 67, where the broken line is a reference spheroid symmetrical about the equator, while the full line shows the profile of the sea-level surface averaged over all longitudes — the average meridional geoid section, to use more scientific language. This is the 'pear-shaped Earth': the 'stem' at the north pole is 18 m high, and the depression at the south pole is 27 m deep. The scale of the diagram is of course exaggerated, and the Earth is not really concave at the south pole. Apart from this necessary distortion, however, the diagram means what it says: if you go to the north pole, dig a hole down to sea level and clamber in, so that you stand at sea level, your feet will be 45 metres further from the equator than the feet of an equally eccentric explorer who bores through 5 km of ice to do the same thing at the south pole. Actually you will be 6356773 m from the Earth's centre, while your mate down south will be 6356728 m from the Earth's centre. Fortunately the satellites can tell us this without the heroic labours of exploration — and much more accurately too: the profile of Fig 67, which takes account of harmonics up to degree 36, should be accurate to better than 1 metre over most of the world.

This kind of orbit analysis relies on optical and radar observations and is a very powerful method for determining the longitude-averaged gravity field, because the continual spinning of the Earth automatically averages its effects on satellite orbits, and because those effects are so large. The gravitational attraction of the equatorial bulge can move the orbital plane in longitude by 500 km per day and so, if the longitude is measured accurate to 0.1 km, the flattening should be obtained with an accuracy of 1 part in 100000 from observations over 20 days. Also the odd zonal harmonics, which alter the geoid profile by only 20 m, produce orbital changes of about 5 km — 250 times greater. So if the orbits are determined accurate to 100 metres, you might expect the profile of Fig 67 to be accurate to 40 cm.

The title of this chapter is too kind to Earth. Anyone seeing Fig 67 may ask whether Earth hath not anything to show more fair, and, sliding quickly from Wordsworth to Shakespeare, may answer, 'chang'd to a worser shape thou canst not be'. This answer is wrong, however: there is worse to come, because Fig 67 is only an average over all longitudes and we must now face the prospect of a knobbly variation with longitude.

Like a potato, too

To expose the shape of the Earth in longitude as well as latitude, we have to introduce a more generalized type of harmonic, dependent on both latitude and longitude, not just latitude like those of Fig 65. The Earth's surface can be regarded as covered by a mosaic of these *tesseral harmonics*, as they are called, after the tessera in Roman pavements. A complete set of tesseral harmonics gives a picture of the ups and downs of the geoid over the whole Earth.

Large numbers of tesseral harmonics are needed to provide a realistic model of the Earth, and calculating their values calls for huge computational efforts and a variety of methods. In recent models more than a million observations, of many types, are used to derive about 1300 harmonics. The data include camera, Doppler and laser observations of satellites in about equal proportions, and measurements of the strength of gravity on Earth and of the height of satellites over the oceans recorded by altimeters. The methods utilize both orbital perturbations and geometrical geodesy.

The outcome of these marathon calculations is a contour map of the geoid relative to a reference spheroid, and Fig 68 shows the map given by what is currently the best model, the Goddard Earth Model 10B, derived in 1978 at the Goddard Space Flight Center near Washington. This map gives the height of the geoid in metres above or below a reference spheroid with an equatorial radius of 6378.140 km and a flattening of 1/298.257. In Fig 68 the elevated areas of the geoid are shown white and the

Fig 68 *Contour map of the height of the geoid in metres relative to a spheroid of flattening 1/298.257, as given by the Goddard Earth Model 10B (GEM 10B). The depressed areas are shaded and the elevated areas white. The contour lines are at 10 m intervals.*

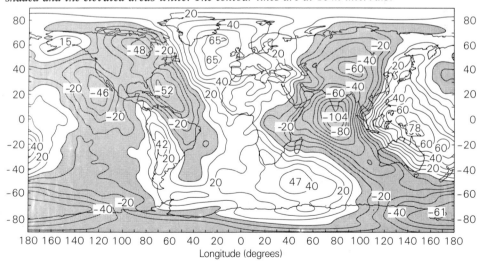

151

depressed areas are shaded. This geoid map should, of course, be consistent with the pear-shaped section of Fig 67 — it *is* the same Earth. The comparison can best be made at the poles, where Fig 68 indicates a north polar hump of nearly 20 m and a south polar depression of about 25 m, in agreement with Fig 67. (The small difference in the flattening of the reference spheroids in Figs 67 and 68 changes the heights by less than 1 metre.)

Fig 68 shows that the greatest departure of the actual geoid (sea-level surface) from the reference spheroid is the depression 104 metres deep, south of India. Not far

Fig 69 *A slice through the equator, showing how the sea-level surface in GEM 10B (unbroken line) departs from a circular shape (broken line). The deepest dip is south of India 100 m deep (part of the 104 m depression in Fig 68) and the highest hump is near Indonesia 72 m high (part of the 78 m hump in Fig 68). The vertical scale is greatly exaggerated.*

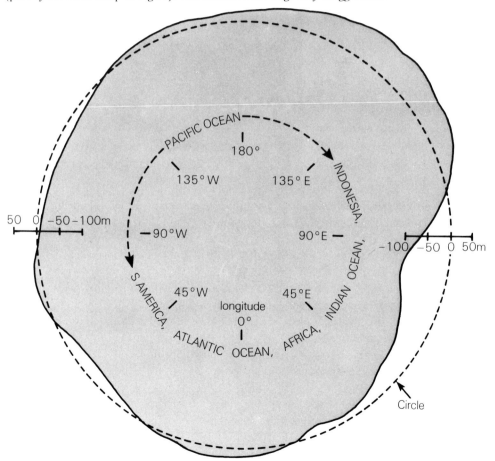

away, near New Guinea, is the highest hump, 78 m above the spheroid. Since the reference equator is exactly circular, this means that if you sailed along the equator from south of India to north of New Guinea (with diversions to avoid land) you would increase your distance from the Earth's centre by about 180 m, although of course you would never have to go uphill, the sea-level surface being 'level' by definition. The two other major humps are near Iceland (65 m high) and south of Madagascar (47 m). The dips are more numerous: the deepest are south of New Zealand (61 m deep) and north of the Himalayas (60 m), though the latter is an offshoot of the depression south of India; the others, which form a triangle straddling North America, are south of Bermuda (52 m deep), in Hudson Bay (48 m deep) and off California (46 m). This geoid map should be accurate to 1 or 2 metres generally, apart from detailed features covering only small areas.

Ordinary maps of the world give the coastline, which is superficially important, because we like to know whether we are on land or at sea; but the geoid map has a deeper significance, because the geoid is the Earth's attempt to express its internal structure, and if only we could read the variations more intelligently, we should learn much about our planet's interior. An ordinary atlas-map is a surface photograph; a geoid map is like an X-ray or body-scanner picture. So the geoid map is worth trying to remember, and a good way of doing so is to focus on the dark (depressed) areas in Fig 68: you see that the western hemisphere is dominated by an animal rather like a goat, confronting (or talking to?) a man with a jutting chin, rather like Popeye, who is in charge of Asia.

This then, is the Earth we live on, and its profile when cut through the equator, as given by GEM 10B, is shown in Fig 69. Since the deepest dip and highest hump on the geoid map are both near the equator, Fig 69 samples them almost in their full glory, showing a 100 m dip at longitude 77°E, south of India, and a 72 m hump at 135°E, north of New Guinea. So if you go and swim in the sea on the equator at 77°E, your head will be 6378040 metres from the Earth's centre, while a mate who goes for a swim at 135°E will be 6378212 metres from the Earth's centre. The best that can be said for this useless experiment is that it is less arduous than standing at sea level at the north pole — unless some monster of the deep digests either of you.

If we go to the opposite extreme in accuracy, we can fairly describe the equatorial section of Fig 69 as rather like a potato, whence the title of this section and the following useful rhyme:

When you cut a slice
Through the polar ice,
    The Earth is like a pear.
But sliced along the equato
She looks like a potato —
    A giant *pomme de terre*.

The virtues of resonance

By now you may be thinking that further advances in knowledge of the Earth's shape will come only from observations of accuracy better than 1 metre, and that visual observations are useless. But that is quite wrong: we have already seen how the longitude-averaged profile can be determined accurate to 40 cm from observations accurate to 100 m, and we can play a similar trick with some of the longitude-dependent harmonics, if we concentrate on *resonant* orbits.

An orbit is said to be resonant if its track over the Earth repeats after a number of revolutions. The most useful example is 15th-order resonance, when the track repeats after 15 revolutions in one day: in other words, the Earth rotates once relative to the satellite's orbital plane while the satellite completes exactly 15 revolutions. This happens when the orbital period of the satellite is close to $\frac{1}{15}$ of a day, that is, 96 minutes; but the exact period for resonance varies according to the orbital inclination, being 95 minutes for 70° inclination and 96 minutes for 100° inclination.

If the ground track repeats after 15 revolutions, successive tracks must be 24° apart in longitude. Now the tesseral harmonics of order 15 correspond to shapes with 15 'petals' round the equator, just as the fifth zonal harmonic has five 'petals' in latitude (Fig 65). That is another way of saying that 15th-order harmonics have 'humps' every 24° in longitude, so that a 15th-order resonant satellite, with tracks spaced at 24°, suffers the *same* perturbation on each revolution from 15th-order harmonics. Thus a tiny perturbation, typically about 1 metre per revolution, builds up 15 times a day, day after day, month after month, into an orbital change that can be accurately measured, to give accurate numerical values for the 15th-order harmonics.

In practice satellites enter resonant orbits only by chance: so far, no launching authority has been kind enough to place satellites deliberately in suitable resonant orbits, though repetitive ground tracks are sometimes required for other purposes, as with the 74° Cosmos satellites that provided excellent examples of 15th-order resonance in the 1970s. However, the average height of a satellite at 15th-order resonance is near 500 km, where air drag is appreciable. The drag steadily reduces the orbital period and, as a result, a satellite starting life with an orbital period of, say, 97 minutes is gradually 'dragged through' the resonance. If the drag is not too great, the effects of the resonance may continue for a year or more, building up to a large value.

Fig 70 shows an example of resonance in action. The circles show 129 values of orbital inclination for Intercosmos 11 (1974-34A) at weekly intervals between 1 June 1975 and 26 November 1977, as given by orbits determined solely from Navspasur observations accurate to about 200 metres. Intercosmos 11 experienced exact 15th-order resonance on 1 October 1976. A theoretical curve has been fitted to the points in Fig 70, with the numerical values of the harmonics chosen to give the best possible fit. It is obvious that the fitting is very good, and the 15th-order harmonics obtained from this and other satellites are accurate to 2%. Fig 70 shows how the effect of the resonance built up steadily between May 1976 and January 1977, and increased the inclination by 0.08°, which is equivalent to 10 km in distance. The total contribution of the 15th-order harmonics to the geoid height is about 50 cm: so the 2% accuracy achieved is equivalent to an accuracy of 1 cm in geoid height — better than even lasers can achieve.

Fig 70 *The orbital inclination of Intercosmos 11 (1974–34A) between 1 June 1975 and 26 November 1977, showing the variations produced by 15th-order resonance, which caused the inclination to increase by 0.08° (equivalent to 10 km) between May 1976 and January 1977. Exact resonance occurred on 1 October 1976. The theoretical curve shown fits the points extremely well, and gives values of lumped harmonics accurate to 2%. (From D. M. C. Walker, Geophys. J. Roy. Ast. Soc., Vol.67, pp.1–18 (1981).)*

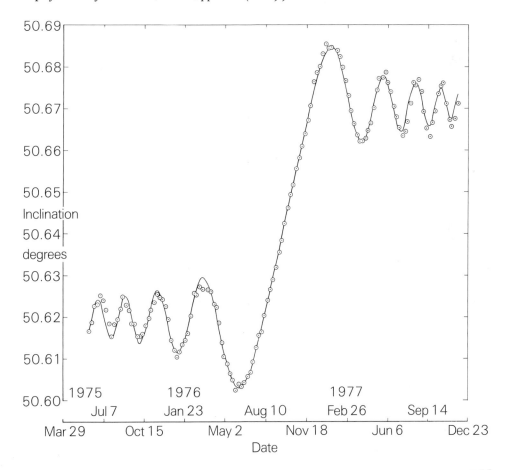

Although comprehensive gravity models such as GEM 10B give a good overall picture of the geoid shape, the accuracy of their 1300 individual coefficients is very difficult to assess, and some sceptics have insinuated that errors of 100% may occur. That is why the values derived from resonance are so important. They are completely independent of GEM 10B and, being much more accurate, provide an exacting test of the accuracy of individual GEM values of order 15. (In fact these GEM 10B values are found to be accurate to about 15%.) Having served this purpose, the coefficients obtained from resonance can then be incorporated in new gravity-field models.

There is nothing special about the 15th-order resonance, and resonances of order 14, 13, 12, ... can also be analysed — indeed the analysis is easier than for 15th-order resonance because the drag is lower. Good results have been obtained for 14th-order (period about 102 minutes), but unfortunately there are very few satellites available for resonances of order 13, 12, 11, ... At these heights, where air drag is scarcely perceptible, existing satellites may not reach resonance for 50 or 100 years, which is too long to ask the analysts to wait. Their best hope is a new satellite launched by chance close to resonance.

In contrast, plenty of satellites experience 16th-order resonance; but this occurs when the period is near 90 minutes, and high drag usually prevents accurate analysis, although successful results have been obtained for one satellite, Skylab 1. Even worse is 17th-order resonance, which requires the satellite to be under the ground! Much better are the 2-day resonances, such as 29:2 and 31:2, when the satellite makes 29 or 31 revolutions in 2 days. The effects are smaller than for the one-day resonances and the results are less accurate, but still useful.

For visual observers, resonance has the great virtue of enabling them to test and improve on the accuracy of geoid maps derived from much more accurate and more expensive observations. Resonance is a magic wand that allows the visual observers and Navspasur to achieve more than is within the power of those who work with specially launched satellites and the latest gimmicks in tracking. We are reminded of the *bon mot* that resonance is the only phenomenon of lasting importance in celestial orbits: all else is merely revolutionary motion, which goes endlessly on but achieves nothing.

To be fair, however, resonance will play only a limited role in improving knowledge of the gravity field, unless someone kindly launches several satellites into orbits at 13th-order resonance, several more at 12th-order resonance, and so on. It is more likely that the main progress will stem from results obtained using accurate modern techniques like lasers, radar altimeters and satellite-to-satellite tracking; but resonance-testing of the resultant gravity-field models will still be needed, to assess their accuracy.

Whatever the method, the aim is the same: it is all part of the campaign 'for improving Natural Knowledge', to use the wording of the second Charter of the Royal Society in 1663. Improving knowledge of the Earth carries with it the hope of understanding the workings of the Earth's crust and interior, and of alleviating the human misery caused by earthquakes.

To end as we began, by quoting from *Macbeth*, we can safely say that deep knowledge about the not so 'sure and firm-set Earth' is being successfully 'plucked from the air'.

10
The Blue
Dome of Air

those streams of upper air
Which whirl the Earth in its diurnal round.

P.B. Shelley, *The Witch of Atlas* (1820)

Having grappled with gravity, we can now leap up with levity and rise balloon-like from the solid Earth into thin air, where until recently 'the owl-winged faculty of calculation dared not ever soar' —

> The upper air
> is very rare.
> We never care
> to go up there.

'It may be rare, but it's everywhere', is a fair riposte to this absurd doggerel, for the upper atmosphere is not merely 'the biggest thing in the world': it is a thousand times bigger than the world. Not so long ago, people looked on the atmosphere as a thin spherical shell of air clinging to the Earth. Now the roles are reversed, and we can look on the Earth as 'an island in the ocean of the air', a tiny nucleus of solid substance in a vast atmosphere extending out to a height of 50000 km on the sunward side and much further on the dark side — an atmosphere shaped rather like a tadpole with its tail away from the Sun.

Not so long ago, too, people regarded the Earth (and its airy shell) as moving through the 'absolute blank' of interplanetary space. Now we see the Earth's vast tadpole-atmosphere as swimming in the even vaster solar atmosphere, which is made up of particles pouring out from the Sun (mostly protons and electrons) at speeds averaging about 400 km per second and extending to beyond the orbit of Neptune.

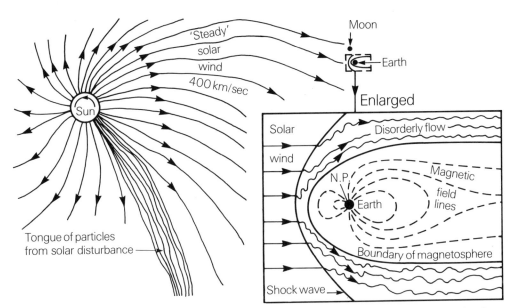

Fig 71 *The Earth in the solar wind. A flow of particles pours out from the Sun at speeds of about 400 km per second. The Earth and its magnetic field form an obstacle in this supersonic flow, and a shock wave develops at a distance of about 90000 km on the Earth's sunward side, as shown in the inset diagram. The magnetosphere, with its boundary at a distance of about 50000 km on the sunward side, can be regarded as the Earth's outermost atmosphere. The inset diagram is a slice through the poles, NP being north pole.*

The Earth's atmosphere, a small obstacle in this all-pervading supersonic outflow, creates a standing shock wave at a distance of about 15 Earth radii on the sunward side, as shown in Fig 71. So we now know that a supersonic airliner has been flying for many million years: it is the Earth itself, moving at 30 km per second, with a sharp shock wave on the sunward side and a long trailing wake on the dark side. The Earth's upper atmosphere, being thus embedded in the Sun's atmosphere, is inevitably under close solar control and responds vigorously when the Sun decides to erupt in a flare and disturb the normal flow of the solar wind. But none of this was known — or even dreamt of — until we observed the first satellites and measured the changes in their orbits caused by air drag.

Slave of the sun

As we saw in Fig 9 on page 17, the effect of air drag on a satellite orbit is to slow the satellite very slightly each time it passes perigee, with the result that the height at apogee steadily decreases and the orbit become more nearly circular. The greater the density of the air at perigee, the more rapidly the orbit will contract. So, if we measure the rate of contraction, we can calculate the air density, provided we know the size, shape and mass of the satellite. This method was first applied in 1957, within two weeks of the launching of Sputnik 1, and the results then obtained have proved to be correct. By the end of 1958 it became apparent that the air density at heights of 200-500 km depended on the activity of the Sun, and we can now recognize several different facets of this solar control.

The greatest changes in air density are produced by the eleven-year cycle of solar activity. As has been known for more than a hundred years, the number and area of sunspots on the face of the Sun undergo a variation which, although irregular, reaches a maximum on average every eleven years and a minimum about six years later. No one has yet been able to predict accurately either the timing or the intensity of the solar cycle. The maximum of activity in 1957–8 was the highest that has ever been recorded, and after a minimum in 1963–5, the next maximum in 1968–70 was

Fig 72 *Variation of air density with height by day and by night, for high and low solar activity. At the solar maximum of 1979–81 the density was between the 'High' and 'Exceptionally high' curves. The scale for density is logarithmic, and the density is a million times greater at the right-hand edge than at the left.*

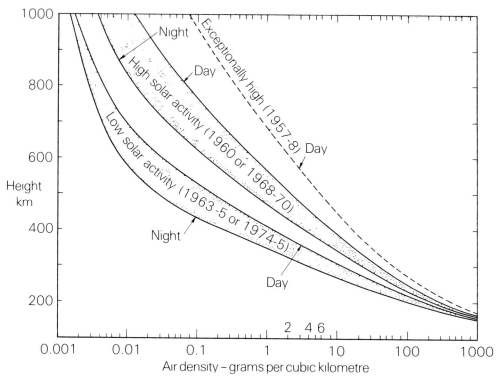

not so intense. After the minimum of 1974–6 the subsequent maximum in 1979–81 was stronger than its predecessor, but still not up to the levels of 1957–8.

The shaded bands in Fig 72 show how the air density varies with height at typical sunspot maxima and minima. The values for 1979–81 were between the 'High solar activity' band and the 'Exceptionally high' curve. The air density in Fig 72, given in grams per cubic kilometre, is on a logarithmic scale and runs from $\frac{1}{1000}$ gram per cubic kilometre on the left-hand side to 1000 grams per cubic kilometre on the right — a factor of a million across the diagram. Fig 72 shows that at heights of 400 – 700 km the density is more than ten times greater at sunspot maximum than at sunspot minimum. At lower heights (and at greater heights) the difference decreases, but is still important down to 200 km, where the density at sunspot maximum is about twice that at sunspot minimum, and up to well above 1000 km, where the density is four times greater at sunspot maximum than at sunspot minimum.

The second solar control over the upper atmosphere is the day-to-night variation in density. As we know from experience on the Earth's surface, the air is usually warmer in the daytime, and the same situation applies in extreme form in the upper atmosphere, where the temperature is hundreds of degrees higher by day. High temperatures go with high densities, and we find that the density undergoes a regular daily cycle, increasing during the morning to a maximum at about 2 or 3 pm in the afternoon and then declining during the evening and night to a minimum at about 4 am. Again the effect is greatest at heights of 400 – 700 km, and Fig 72 shows that the maximum daytime density is about five times greater than the minimum night-time density at a height of 500 km. The maximum day-to-night variation occurs at about 400 km height when solar activity is low and at about 600 km when solar activity is high. Every day, whether it rains or shines at sea level, the upper atmosphere undergoes this huge variation: the air 500 km above you will be (or was) about five times denser at 2 pm this afternoon than it will be at 4 am tomorrow morning.

Fig 72 shows that it is no good asking, 'What is the air density at 500 km height?' unless you specify the solar activity and time of day. If it is 4 am on a winter morning in 1985, when solar activity will be low, the density will be about 0.03 grams per cubic kilometre. But at 3 pm on a summer afternoon in 1990, the density will be about 3 grams per cubic kilometre — a hundred times greater — if solar activity rises to an average maximum by 1990.

Fig 72 gives an impression of the density on an average day; but this is really an over-simplification, because most days are non-average. The Sun, especially during sunspot maximum and for a year or two afterwards, is scarcely ever quiescent and is continually producing minor (or major) eruptions which inject streams of higher-energy particles into the regular flow of the solar wind. To add to the confusion, the Earth may fly smack into one of these streams; or just sample its edges; or miss it altogether. Major solar outbursts, which disrupt the solar wind, are always felt by the atmosphere: if the Earth 'receives a basinful', as it were, the density at 600 km

161

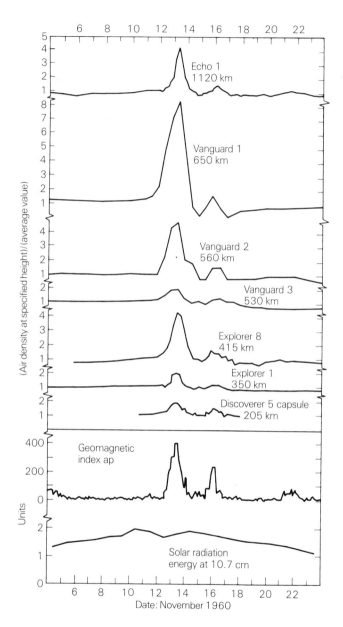

Fig 73 *The departure of upper-atmosphere density from its average value at heights between 200 and 1200 km, as revealed by the orbits of seven satellites during November 1960. The changes in density closely parallel the changes in the geomagnetic index. (After L. G. Jacchia,* Space Research II, *p.747.)*

may increase to eight times its previous level and the upper-atmosphere temperature may rise by 500°. Fig 73 shows the variations in density at various heights as calculated by Luigi Jacchia at the time of the great solar disturbances in November 1960. The geomagnetic index *ap* is a measure of the disturbance of the Earth's ground-level magnetic field, which is often used as a measure of intensity of the solar disturbance. From 4 to 12 November the density at all heights is fairly constant; then on 13 November there is a sudden large increase in air density, exactly in unison with the geomagnetic index in the lower part of the diagram. Two days later comes a smaller solar disturbance, also faithfully reflected in the upper-atmosphere density at all levels. Fig 73 also shows that 650 km (Vanguard 1) is the height at which the atmosphere is most sensitive to the solar disturbances.

Though such a large disturbance is rare, smaller ones often arise, sometimes once a week, sometimes once a day — and sometimes not at all for weeks if the Sun is at its quietest. Consequently the air density varies rather irregularly from day to day, by up to about 20%, and these day-to-day twitches are largely unpredictable, because we do not know exactly how each solar disturbance will affect the solar wind or whether it will impinge on the Earth; and, when there is an effect on the Earth, it may be concentrated in northern or southern latitudes, or may spread over the whole Earth.

You may be surprised that the Sun exerts such power over the upper atmosphere, but it is to be expected because the energy released is so great. Visually, the most spectacular results of solar disturbances are the displays of the aurora, produced when the population of charged particles in the polar regions of the Earth is suddenly augmented by an influx from the solar wind. The excess is 'dumped' into the upper atmosphere, raising the temperature and producing the beautiful shimmering coloured curtains of the aurora. The energy released in a bright aurora can be as much as 10 million megawatts, which is more than the total world electricity supply, and even a moderate aurora can outshine all the bright lights of America, as shown in Fig 74, which is a composite picture made on several successive passes of a meteorological satellite. The photograph shows an aurora over Canada and the city lights of the USA beneath.

An auroral display is usually a sign that the upper atmosphere is being heated strongly. But the heat does not stay put in the polar regions: it propagates round the world and affects the temperature and density at other latitudes in a variable way at variable times. The 'average atmosphere' of Fig 72 then becomes rather a myth: the upper atmosphere, like the ocean, is rarely calm and often very rough indeed, as in Fig 73. Rough weather near the ground can also add to the *mêlée* in the upper atmosphere. Some of the energy in a thunderstorm may find its way upwards, where it has a dramatic effect because the air is so rarefied. These waves from 'down under' are another source of variability.

Satellite orbits are not too fussy about the make-up of the air they move through.

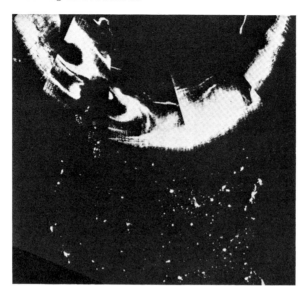

Fig 74 *Composite photograph of a moderate aurora over Canada and the city lights of the USA to the south, taken on 14 February 1972. The lights of the Washington–New York–Boston area can be seen on the right, with the Florida coast at the lower right. On the left side San Francisco and Los Angeles stand out, and the largest area of light in the centre is around Chicago.*

They react to its density and are only slightly affected by changes in its composition. At sea level the air is about 80% molecular nitrogen and 20% molecular oxygen, and this composition is maintained more or less unchanged up to a height of 200 km. Then oxygen, split into atomic form, becomes dominant from 200 km up to about 600 km, for medium solar activity. Above that the lighter gas helium dominates from 600 to 1800 km, and at the uppermost levels we find the lightest gas of all, atomic hydrogen, in command. So the intuitive idea of lighter gases rising to greater heights has proved correct. Erasmus Darwin expressed it neatly in 1792 when he referred to 'brighter regions',

> Where lighter gases, circumfused on high,
> Form the vast concave of exterior sky.

As might be expected, the upper and lower limits of the helium zone depend on solar activity: for low solar activity, helium is dominant from 500 to 800 km only; for medium solar activity, from 600 to 1800 km, as already indicated; and for high solar activity, from 800 to well over 2000 km.

I have deliberately left until last a very important atmospheric variation which is only indirectly linked with the Sun. This is known as the semi-annual variation because it has a six-monthly rhythm. Every year we find that, quite independently of any variations in solar activity, the air density throughout the upper atmosphere is low in January, rises to a maximum in late March or early April, then falls to a deep minimum in late July or early August, and rises to another maximum in late October or early November, before falling off again towards the January minimum. The July

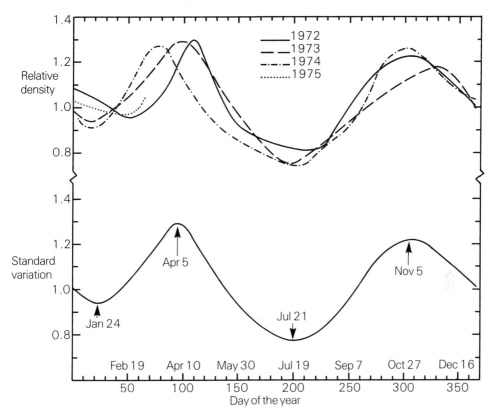

Fig 75 *The semi-annual variation in density during the course of the year at heights near 250 km during the years 1972–75, after removal of day-to-night and solar activity effects. The semi-annual variation itself varies from year to year and the curve below gives an averaged version. From analysis of Cosmos 462 by D. M. C. Walker (Planetary and Space Science, Vol.26, pp.291–309 (1978).)*

minimum is always lower than the January minimum, but the April and October maxima are approximately equal — the 'winner' changes from year to year. The strength of the semi-annual effect and the dates of the maxima and minima also vary from year to year.

Fig 75 shows the variations during the years 1972-1975 at heights near 250 km, from a classic study by Doreen Walker using 604 values of density obtained from analysis of visual, optical and Navspasur observations of Cosmos 462 (1971-106A). The actual variation of density is distinctly jagged, and the curves are drawn through the ups and downs to reveal the underlying trends. Fig 75 clearly displays the variations in strength from year to year and the shifts in the dates of the maxima and minima, and also gives an averaged 'standard density variation' for 1972-1975.

The semi-annual variation depends on height, and is strongest at heights near 500 km, like the solar-activity and day-to-night effects. At 500 km the maximum density in April or October may be as much as 5 times the minimum in July. This factor decreases to about 2 on going up to 1000 km, and also decreases on going down to 250 km, where Fig 75 shows that it is about 1.8. But the semi-annual variation is special in that it remains strong below 200 km and is the most important of all the variations at 150 km, where the effects of solar activity are weak.

The semi-annual variation seems to be linked with changes in the lower atmosphere, and although the exact mechanism is disputed, two facts seem sure to be relevant. First, the main six-monthly rhythm probably reflects seasonal changes in the lower atmosphere, and particularly the fact that the temperature pattern at equinox (March-April and September-October) is different from that at solstice (December-January and June-July). The maxima and minima of the semi-annual variation occur about a month after equinox or solstice. Second, the density being lower in July than in January reflects the fact that the Earth's distance from the Sun is 3% greater in July than in January, so that worldwide we receive 6% less solar radiation in July than in January. The semi-annual variation qualifies as 'a slave of the Sun' because of the solar-distance effects and because it almost certainly has its origin in the lower atmosphere, which itself dances to a solar tune.

Most of our air-density satellites, like the actors in Prospero's 'insubstantial pageant' in *The Tempest*, have 'melted into air, into thin air', having burnt up in the thin air of the high atmosphere after living out a life of useful scientific service in the even higher atmosphere, where

> the winds and sunbeams with their convex gleams
> Build up the blue dome of air,

in Shelley's words. By reacting so sensitively to the tenuous grasp of the rarefied air, satellite orbits have unveiled many secrets of this aery dome, though some of its interwoven variations still conspire to perplex us.

The viewless winds

Winds at ground level are generated by the differences in air pressure around depressions and anticyclones. In the upper atmosphere there are no such local weather systems; but there is a great difference in pressure between day and night, the maximum daytime pressure being up to 5 times greater than the minimum night-time at heights of 300-700 km, just as with the density. So strong winds can be expected.

'How on earth can satellite observations help to measure the winds?' you may ask. The answer is, 'by revealing the decrease in the inclination of the orbit to the equator'. If the atmosphere did not rotate, air drag would act in the plane of the orbit and the orbital inclination would remain unchanged. But the rotation of the atmosphere subjects the satellite to a sideways force, which has the effect of slightly reducing the orbital inclination as the satellite's life progresses. The faster the atmospheric rotation, the greater the decrease in inclination: so, by accurately measuring the decrease, we can determine the atmospheric rotation rate.

It is convenient to measure the atmospheric rotation rate in revolutions per day (rev/day), so that a rate of 1.0 rev/day implies an atmosphere rotating at the same rate as the Earth, with the wind speed zero. A rotation rate of 1.2 rev/day corresponds to a west-to-east wind of 0.2 times the rotational speed; and at 200 km height the rotational speed is about 480 metres per second (m/s) on the equator, decreasing as latitude increases, to 340 m/s at 45° latitude. So if a satellite in an eccentric orbit with perigee at the equator indicates that the rotation rate is 1.2 rev/day, the wind speed is about 96 m/s west-to-east near the equator.

In practice, atmospheric rotation produces only a small decrease in orbital inclination. A circular polar orbit in an atmosphere rotating at 1.0 rev/day suffers a decrease of 0.007° in inclination for each decrease of 1 minute in orbital period. So if the orbital

Fig 76 *Daily values of orbital inclination for Heos 2 second stage (1972–05B) at the end of its life, with theoretical curves for atmospheric rotation rates of 1.0 and 1.4 revolutions per day. The latter obviously fits better. The average perigee height was 240 km and the local time at perigee between 1 am and 2 am. (After H. Hiller,* Planetary and Space Science, *Vol.29, pp.579–588 (1981).)*

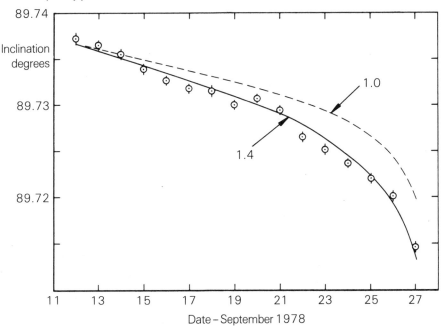

period decreases by 2 minutes, from say 97 to 95 minutes, the inclination would decrease by 0.014°. If the measured decrease differed from this, and was found to be, say, 0.017°, that would imply an atmospheric rotation rate of 1.2 rev/day; and since the 'averaged' latitude for a polar orbit is about 30°, this indicates a west-to-east wind of about 80 m/s.

Fig 76 shows the method in action: the diagram gives the daily values of inclination for the second-stage rocket of Heos 2 (1972-05B) in the last 14 days of its life, as determined mainly from NORAD radar observations. Also shown are two theoretical curves, for atmospheric rotation rates of 1.0 and 1.4 rev/day, and it is quite obvious that the 1.4 curve is better. This implies an average west-to-east wind of about 150 m/s near the perigee of the satellite, which was at local time between 1 am and 2 am and a height of 240 km. Many such values of rotation rate have been obtained over the years from a variety of satellites. Few of the results are as clear-cut as Fig 76 but, taken together, they tell us how the atmospheric rotation rate varies with height and local time.

Fig 77 gives a picture of the results, which have been divided into three categories — 'morning' (approximately 4 am to noon local time); 'evening' (approximately 6 pm to midnight local time); and 'average', covering results from circular orbits and also from non-circular orbits where perigee experiences a wide range of local times without much bias. These categories are only roughly defined: for example, the end

Fig 77 *Upper-atmosphere rotation rate (in revolutions per day) with corresponding wind speeds (in metres per second, on right). The full line represents average conditions, the broken lines morning and evening. The individual values are from many different satellites and the vertical bars indicate the likely errors of the values. (After D. G. King-Hele and D. M. C. Walker,* Planetary and Space Science, *Vol.25, pp.313–336 (1977).)*

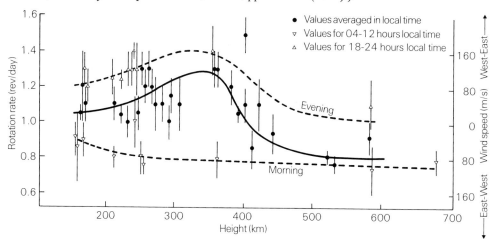

of the 'evening' category is stretched a little to catch 1972-05B. Despite the approximations, the results are still illuminating.

In Fig 77 the left-hand scale gives the rotation rate, and the right-hand scale gives the corresponding wind speed, taken to be at the 'average' latitude of 30°.

The unbroken curve in Fig 77 shows that the average rotation rate increases from 1.0 rev/day at a height of 200 km to a maximum of 1.3 rev/day at 350 km, and then seems to decrease to 0.8 rev/day by 600 km. At heights between 200 and 400 km the atmosphere is going round faster than the Earth on average, or 'super-rotating', as it is sometimes called. The quotation from Shelley at the start of this chapter is therefore unexpectedly apt. In terms of west-to-east winds, the curve of average rotation rate in Fig 77 gives a near-zero average wind at 200 km, 120 m/s west to east at 350 km and 80 m/s east to west at 600 km.

The upper broken curve in Fig 77 shows that the west-to-east winds are strongest in the evening — about 80 m/s at a height of 200 km, increasing to about 150 m/s at 350 km, and then declining. In the morning the winds are generally from east to west. Of course, Fig 77 needs strengthening with many more values, and there are some discrepancies, probably because the violent changes in density and pressure at the time of solar storms create ferocious winds with speeds of up to 1 km per second. The 'averaged' results in Fig 77 are inevitably a little distorted by such extremes.

North-to-south winds can also sometimes be determined from changes in orbits, but as yet there are few good results.

Carry on observing

Much of our recent knowledge of the upper atmosphere has come from analysing satellite orbits determined from observations. But there are many new and sophisticated instruments being developed to measure the properties of the upper atmosphere on the spot from spacecraft such as the Shuttle. Will the observations — visual, optical and radar — still be needed in the future?

The answer is definitely 'Yes'. To begin with, determining atmospheric densities and winds from observations is by far the cheapest method, and does not require either the development of special instruments or the launch of a satellite for the purpose. Secondly, observation and orbit analysis gives the density variations more accurately than other methods: 2% accuracy can readily be achieved, and 1% by taking more trouble. Thirdly, observation and orbit analysis produces results over many years with no deterioration in accuracy. This is just what is needed for measuring the variations in density over the eleven-year solar cycle, and the year-to-year vagaries of the semi-annual variation. Instruments aboard satellites cannot compete because of problems with cost, power supply, reliability and calibration accuracy over say ten or twenty years.

Observations of satellites in fairly low orbits by visual or radar methods are ideal for air density researches, because visual and radar tracking is tolerant of the likely prediction errors of up to 2 or 3 minutes and because the orbits determined from these observations have just about the right accuracy — 100 or 200 metres in height. At heights of 200-300 km a change of 1 km in height alters the density by about 3%, so an orbital error of 200 metres creates an error in density of about $\frac{1}{2}$%, which is quite acceptable because errors of 1% arise from other sources.

The timing accuracy of visual observations, 0.1 second, may seem poor beside the marvels of modern microelectronics, but it is good enough to determine air density. Two such observations a day apart give the orbital period of a satellite accurate to 0.01 second, if it makes about 14 revolutions per day. A typical 'air-density satellite' like Cosmos 462 (used in Fig 75), suffers a decrease in orbital period of about 1 second per day. So the visual observations allow the daily rate of decay to be determined accurate to about 1%. (If the decay rate is slower, a similar accuracy can be achieved over a longer time interval.) All in all, the observations are well matched to the requirements.

Visual and radar observations alone, however, are not usually accurate enough for measuring atmospheric rotation rates over short time intervals, and more accurate observations such as those from Hewitt cameras are really needed. A good 'atmospheric-rotation satellite' will undergo during its life a decrease in inclination of about 0.1°, which is equivalent to a decrease of 11 km in the maximum latitude it reaches. Also it needs to be in orbit for about a year if enough observations are to be made, and the average decrease in maximum latitude would then be equivalent to 900 metres per month. To obtain monthly values accurate to 5%, we need orbits accurate to 50 m or better, which is possible if visual and radar observations accurate to 200 m are used in conjunction with Hewitt camera observations accurate to 10 m.

There are other methods for determining winds, such as radar backscatter, and measuring the motion of vapour trails or the Doppler shifts of airglow spectra. But these are essentially measurements of local winds and do not give the overall rotation rate. Measuring wind velocity by instruments aboard satellites has so far proved expensive and rather inaccurate. But an accurate satellite-borne wind gauge could relegate the observational method to a back seat. For the foreseeable future, however, the study of rotation rates by analysis of orbits determined from observations will remain a thriving area of research.

Russell Eberst likes to look on satellite observing and analysis as a plant, with predictions as the roots, observations as the main stem and research results as the flowers. These last two chapters show how the flowers are continuing to bloom through many a summer.

11
Into Eclipse . . .

The mortal moon hath her eclipse endured.

William Shakespeare, *Sonnet* (c.1595)

The mortal moon Vanguard 1 has endured eclipse more than ninety thousand times since its launch in 1958, and enjoyed more than ninety thousand sunrises. Before it was launched, scornful journalists called Vanguard 1 'the grapefruit satellite', because it was a sphere only 15 cm in diameter weighing less than 2 kg and carrying only a radio transmitter. But the scoffers were merely showing their ignorance, for Vanguard 1 has enjoyed a gloriously fruitful career. It was the first satellite to use solar cells for power. Its orbit was the first to show clearly the day-to-night variation in air density, and was the only orbit to sample the atmosphere above 600 km in 1958 after the highest-ever solar activity. The changes in perigee height revealed the Earth's tendency towards a pear shape; and, being the oldest, it will remain the most valuable orbit for detecting changes in the Earth's shape as the years go by. So presumably it is being observed intensively, you may think. Wrong! It is only being observed by radar and by a few visual observers with large telescopes, because the Baker-Nunn cameras, which kept watch on Vanguard 1 for so many years, now only operate in support of laser tracking.

In our wicked world, showy trash eclipses long-term worth, and the fate of Vanguard 1 is one small example. Before my narrative itself passes into eclipse, I ought to ask how the gentle art of satellite observing will fare in a world where the nasties make all the running, grabbing as much as possible for themselves and wielding power over the despised 'do-gooders', who would like to maximize human happiness, physical and intellectual. My sympathies are with the latter group, and I have always hoped that science would help people to a better life. In recent years this has not happened (except perhaps with microelectronics), with the result that science and society have become alienated, and people eye critically all scientific expenditure, asking such questions as 'why spend money on launching satellites?' This question is misguided, because 70% of space expenditure is military and, if space launchings were banned, it would cost more to perform the same military functions — reconnaissance, navigation, etc — by other methods. So no money would be saved. If the question really means diverting all military expenditure (including the tiny fraction on space) to improve human happiness, that is a splendid idea — but a mere fantasy in the world as we know it, where rulers provoke childish confrontations with other nation-states to divert attention from their own misdeeds.

So space launchings are likely to continue, and the real question is, should we let the satellites and rockets go unobserved and unused once they are in orbit? To that I would answer, 'No'. I think we should try to get as much 'value for money' as possible out of them — especially if it was someone else's money that launched them — by observing them and analysing their orbits for researches on the upper atmosphere and gravitational field. The military satellites alone would be enough: for example, the three years of data on the semi-annual variation of density in Fig 75 were derived from the orbit of Cosmos 462, a 'killer' satellite. Whatever the purpose of their launching, all the objects in orbit, whether gold-plated satellites or discarded junk, are most sensitive probes of the Earth's inner structure and outer atmosphere, as the two previous chapters have shown.

Visual observations by volunteer observers not only contribute to the researches but also deserve support for other reasons. The first of their virtues is thrift: the only cost is in distributing predictions and collecting the observations afterwards — no expensive new gadgets, only the 'postage and packing', as it were. The second virtue is social: those members of society who volunteer to observe can find themselves drawn into an absorbing activity that offers unlimited challenges, develops qualities of precision, ingenuity and quick decision, and is satisfying because it contributes to the advance of 'natural knowledge'. Third, and to my mind most important of all, satellite observing involves people in scientific work: this is essential if the alienation of science from society is to be healed, and I only wish the observers were numbered in thousands rather than hundreds. It is a difficult art, but I hope more will attempt it. That is why I have undertaken the hard labour of writing this book.

Internationally, satellite observing and orbit analysis also have great virtue, in promoting goodwill. Near-polar satellites pass over all countries and, to determine their orbits well, we need observations from all over the world. So we have good reason to cooperate, rather than indulge in confrontation, with other nations; and in fact there is excellent international cooperation in the sphere of satellite observations and orbit analysis, organized through the international Committee on Space Research, COSPAR.

What the future will bring is increasingly uncertain; what is increasingly certain is that a future rife with confrontations will bring an end to civilization. Amid this uncertainty we can be sure that observing Earth satellites does at least help to promote the mutual understanding between nations that is needed to save mortals from enduring enduring eclipse.

Appendix

A choice of satellites for observing

Since different people have different preferences, the choice is difficult to make. Also note that future changes in plans may invalidate some of the entries. The list is a mixture of (1) short-lived 'targets of opportunity' like the Shuttle orbiter, bright but sometimes unpredictable because of manoeuvres; (2) reliable long-lived objects; and (3) satellites in eccentric orbits, slow and faint at apogee, but fast and bright at perigee. The column 'St.mag.' gives the standard magnitude at 1000 km distance. The comments are written in 1982, and some may be invalidated by future developments. For example, the 'Big Bird' flights may cease; the Shuttle and Cosmos reconnaissance satellites may stay up longer, or have different inclinations or heights; etc.

Name and designation	Inclination degrees	Perigee height km	Apogee height km	St. mag.	Comments
Shuttle orbiter (STS)	40–57	usually 200–300	usually 300–500	1	Frequent launches. Usually in orbit for about two weeks. Manoeuvrable.
Salyut (with Soyuz, Progress, etc.)	51	200–300	200–300	1	In orbit for several years. Manoeuvrable. Joins with other satellites as space station.
Cosmos reconnaissance satellites	63–82	200–250	250–350	3	30 launches per year. Usually in orbit about two weeks. Manoeuvrable.
LASP 'Big Bird'	96	200–300	300–500	2	In orbit for 6 months-2 years. Manoeuvrable.
Seasat 1978-64A	108	800	800	3	Reliable object. Low drag. Bright. Well predicted.
Agena rocket eg 1964-01A	70	900	900	4	Typical of many Agenas in near-polar orbits. Low drag. Well predicted. Mag.4.

Cosmos 379 1970-99A	51	180	3000 falling	4	Excellent eccentric-orbit objects. Mag.2 at perigee, faint at apogee. Decay about 1984 and 1990 respectively.
Cosmos 398 1971-16A	51	210	6000 falling		
Cosmos 1345 rocket 1982-26B	74	500	540	4	Typical Cosmos rocket. 7.4 m long, 2.4 m diameter. Will decay about 1988.
Cosmos 185 rocket 1967-104B	64	450	800	4	Lower-drag Cosmos rocket. Easy to observe.
Meteor satellites and rockets	81	550–650 or 800–900	650–800 or 850–950	5	Including some called 'Cosmos', over 100 are in orbit. Visible to naked eye when overhead.
Pageos fragments eg 1966-56S	85	about 3000	about 5000	1	Sporting objects. Several fragments of this balloon visible in 1982. But may disintegrate.

Further Reading

Chapter 1

M. Minnaert, *Light and Colour in the Open Air* (Bell, London, 1959).
The beauties and illusions of the sky—haloes, rainbows, autokinesis, etc.

P.J. Klass, *UFOs Explained* (Random House, New York, 1974).
A sceptical survey of sightings.

Chapters 2 and 3

R. Turnill, *The Observer's Spaceflight Directory* (Warne, London, 1978).
Illustrated dictionary of spacecraft, at a reasonable price.

K. Gatland, *Illustrated Encyclopedia of Space Technology* (Salamander, London, 1980).
Lavishly illustrated history of space exploration.

D. G. King-Hele, J. A. Pilkington, H. Hiller and D. M. C. Walker, *The RAE Table of Earth Satellites, 1957–1980* (Macmillan Press, London, 1981).
Complete listing of all spacecraft and rockets, with lifetimes, sizes, shapes and orbits: 672 pages; 2145 launches.

C. S. Sheldon, *Soviet Space Programs, 1971–5* (US Govt. Printing Office, Washington, DC, 1976).
An authoritative survey of the USSR space programme: 668 pages.

N. Johnson, *Handbook of Soviet Manned Space Flight* (Univelt Inc., San Diego, Calif., 1980).
A good review of the flights, including Soyuz and Salyut: 461 pages.

United States Civilian Space Programs, 1958–1978 (US Govt. Printing Office, Washington, DC, 1981).
Covers all aspects, and includes a 'master log' of US launchings, civil and military: 1339 pages.

D. Baker, *The History of Manned Space Flight* (New Cavendish Books, London, 1981).
Detailed account of the US programme.

D. G. King-Hele, 'Methods for predicting satellite orbital lifetimes'. *Journal of the British Interplanetary Society*, Vol. 31, pp. 181–96 (1978).

T. Riggert, 'Skylab's fiery finish'. *National Geographic*, Vol. 156, pp. 581–4 (1979).
Mathematical methods for predicting decay; and description of an actual decay.

For information on current space launchings, see journals such as *Aviation Week* and *New Scientist* (weekly) and *Spaceflight* (monthly).

Chapters 4 and 5

H. Miles (editor), *Artificial Satellite Observing* (Faber, London, 1974).
Chapters 4–6 are particularly useful for visual observers.

J. Heywood (editor), *Artificial Earth Satellites* (British Astronomical Association, Hounslow, 1961).
Particularly the sections by R. Eberst, K. Fea and G. E. Taylor.

A. Bečvář, *Atlas Eclipticalis 1950.0; Atlas Borealis 1950.0; Atlas Australis 1950.0* (Československe Akademie Věd, Prague, 1958; 1962; 1964).
Essential atlases, obtainable from Sky Publishing Corporation, Cambridge, Mass. 02238, USA.

A. P. Norton, *A Star Atlas and Reference Handbook* (Gall and Inglis, Edinburgh, 17th ed., 1978).
The basic star atlas.

Chapter 6

R. H. Chambers, 'An experiment in determining the accuracy of visual estimates of artificial satellite positions'. *Journal of the British Astronomical Association*, Vol. 80, pp. 361–7 (1970).
'Moving-light' experiments on the accuracy of visual observations.

G. J. Kirby, 'The accuracy of visual observations of Earth satellites'. *Quarterly Journal of the Royal Astronomical Society*, Vol. 22, pp. 28–39 (1981).
Assesses the best possible accuracy of visual observations.

J. B. Sidgwick, *Amateur Astronomer's Handbook*, (Enslow, Hillside, New Jersey, 4th ed., 1980). (Faber, London, 2nd ed., 1956).
The optical aspects of astronomical observing.

The Smithsonian Astrophysical Observatory Star Catalog. 4 vols. (Smithsonian Institution, Washington, DC, 1966).
Positions and proper motions of 258997 stars of visual magnitude brighter than 10.

Journals on astronomical observing: *Sky and Telescope* (monthly); *Journal of the British Astronomical Association* (bimonthly); *Popular Astronomy* (quarterly).

Chapter 7

E. N. Hayes, *Trackers of the Skies* (Doyle Publ. Co., Cambridge, Mass., 1968).
Describes the Smithsonian tracking work, with Baker-Nunn cameras and Moonwatch.

J. Hewitt, 'An f1 field-flattened Schmidt system for precision measurement of satellite positions'. *Photographic Science and Engineering*, Vol. 9, pp. 10–19 (1965).
Details of the Hewitt satellite camera.

G. Veis, 'Optical tracking of artificial satellites'. *Space Science Reviews*, Vol. 2 pp. 250–96 (1963); A. G. Massevitch and A. M. Losinsky, 'Photographic tracking of artificial satellites'. *Space Science Reviews*, Vol. 11, pp. 308–340 (1970).
Useful surveys of optical tracking with cameras.

Sky and Telescope, May 1982, pp. 469–473.
Description of GEODSS.

Philosophical Transactions of the Royal Society A, Vol. 284, pp. 419–623 (1977).
Proceedings of discussion meeting on laser ranging to satellites and the Moon.

Chapter 8

G. Perry, 'Radio tracking of satellites', pp. 122–142 in *Artificial Satellite Observing* (ed. H. Miles) (Faber, London, 1974).
An excellent introduction for the expert amateur.

R. L. Easton, 'The Mark II Minitrack System'. *Journal of the British Interplanetary Society*, Vol. 16, pp. 390–419 (1958).
Detailed description of the electronics.

Satellite Doppler Tracking and its Geodetic Applications (Royal Society, London, 1980).
Covers Doppler tracking and Navstar. 196 pages.

'Man-made objects in space', *Spaceflight*, Vol. 19, pp. 410–413 (1977); and in articles in *Aviation Week*, 16 June 1980, pp. 233–42, and 13 April 1981, pp. 68–73.
Descriptions of the NORAD space detection and tracking system.

W. C. Schneider and A. A. Garman, 'Tracking and data relay satellite system'. *Acta Astronautica*, Vol. 8, pp. 179–94 (1981).

Chapters 9 and 10

D. G. King-Hele, 'A view of Earth and Air', *Philosophical Transactions of the Royal Society* A, Vol. 278, pp. 67–109 (1975).
Covers the same topics as the chapters, with more detail.

CIRA 1972 (COSPAR International Reference
   Atmosphere 1972], (Akademie-Verlag, Berlin,
   1972).
   Used as a standard in upper-atmosphere
   research.

Chapter 11
D. G. King-Hele, *The End of the Twentieth
   Century?* (Macmillan, London, 1970).

Organizations to join (in Britain)
The British Interplanetary Society (27 South
   Lambeth Road, London, SW8 1SZ) is for those
   interested in all aspects of space technology
   and science, and publishes two monthly
   journals, *Spaceflight* and *Journal of the British
   Interplanetary Society.*

The British Astronomical Association
   (Burlington House, Piccadilly, London, W1V
   0NL) is the chief society for amateur
   astronomers and publishes the bimonthly
   Journal of the British Astronomical Association.

The Junior Astronomical Society (enquire at
   BAA for Secretary's address) has similar aims
   to the BAA, but with a youthful bias, and
   publishes *Popular Astronomy* quarterly.

The Royal Astronomical Society (Burlington
   House, Piccadilly, London, W1V 0NL) is
   primarily for professional astronomers and
   geophysicists. Its *Quarterly Journal* has many
   articles of general interest.

Picture Acknowledgements
Numbers refer to pages.

Morris Allan 3. Associated Press, London 10
(inset), 21, 37. Astronomical Council of the USSR
120. Crown Copyright Reserved, reproduced by
permission of the Controller, HM Stationery
Office 119. Earth Satellite Research Unit,
University of Aston, Birmingham 117. NASA 11,
31. NASA Goddard Space Flight Centre 125.
NASA/Space Frontiers 13, 36, 44. Novosti 10,
33. Reproduced by permission of Dr E.H. Rogers
164. Royal Greenwich Observatory,
Herstmonceux 126. Royal Observatory,
Edinburgh 123. Smithsonian Astrophysical
Observatory 114. Alf Sorbello, Perth, Australia
23. TASS 40. US Air Force Photo 121. US Navy
141, 142.

For help in obtaining photographs Desmond
King-Hele thanks Dr C.J. Brookes (Fig 49), Flt. Lt
A.A. Dale (Fig 61), Dr H. Hasenfus (Figs 63 and
64), Dr G. Keating (Fig 17), Dr A.G. Massevitch
(Fig 52), Mr G.E. Perry (Fig 12), Dr D.E. Smith
(Fig 55) and Dr G.A. Wilkins (Fig 56). For advice
on various topics in the text we thank Mr
Russell Eberst and Dr Doreen Walker.

Index

178